Exotic
Orchids
IN AUSTRALIA

David L. Jones

Principal photography by
Ron Tunstall

Additional photography by
Gerald McCraith and Mark Clements

Reed New Holland

Acknowledgements

I express my appreciation to a number of people who helped in various ways with this book. Firstly my wife Barbara for her usually competent conversion of my scrawl into a readable manuscript. I would also like to thank Gillian Savage for reading the manuscript and Mark Clements for discussions about various species. Special thanks are due to Ron Tunstall who took the majority of photographs used in the book. Additional photographs were obtained from Gerald McCraith and Mark Clements. Numerous people allowed Ron Tunstall access to their collections and all are thanked sincerely. They include Russell Dixon, Leo Cady, Rick Seabrook, Werner Diesel, Jim Tucker, Brian Whitehead, Mark Webb, Russell Lea, Arnold Faulkner, John Hynds, Anton Van Bysterveld, Bob Dodds, Wal Upton, York Meridith, Graeme Bradburn, Ron Wheeldon, Jack Janese, Bob Trevena, Roger Kramer and the members of the Species Orchid and Carnivorous Plant Society of the Illawarra.

A Reed New Holland book
Published in 1998 by
New Holland Publishers Pty Ltd
Sydney • London • Cape Town • Singapore

First published 1990 by Reed Books, Australia

Produced and published in Australia by
New Holland Publishers Pty Ltd
3/2 Aquatic Drive, Frenchs Forest
NSW 2086 Australia

24 Nutford Place
London W1H 6DQ
United Kingdom

80 McKenzie Street
Cape Town 8001
South Africa

Copyright text © David Jones 1989

National Library of Australia
   cataloguing-in-publication data:

Jones, David L. (David Lloyd), 1944–
   Exotic orchids in Australia.

   Includes index.

   I. Orchids – Australia. I. Title

584'.15'0994

ISBN 1 87633 403 7

Edited by Pamela Polglase
Designed by Cathy Hoare
Cover design Luisa Laino
Typeset by Eversize Typeart Services, Sydney
Printed and bound in Singapore by Imago

Exotic
Orchids
IN AUSTRALIA

Contents

Preface

Of all the groups in the Plant Kingdom, orchids are easily the most popular for cultivation and they have a large following of devout enthusiasts in most countries of the world. In Australia they have almost a cult following the size of which can be demonstrated by the fact that throughout the country more than sixty societies have been formed to coordinate or promote aspects of their cultivation. Some of these are concerned with man-made hybrids whereas others are devoted entirely to species.

I have been associated with the cultivation of orchids since my early teens and I am still a devoted disciple of this magical group of plants. This book, borne out of a desire to share some of that magic, is designed to facilitate orchid cultivation by providing information about the habitats and climate of the areas from which they originate, in conjunction with specific cultural details gleaned from experience. In all some two hundred and thirty exotic species are discussed and illustrated in colour.

Modern orchid literature has been consulted in order to keep abreast of orchid nomenclature and the names used in this book are those of modern botanical authorities. Recent synonyms of the species illustrated in this book or those which are still in common use are included in the index. For accuracy in the application of the botanical name, author citations are provided. Only well known common names are used. A glossary is provided to assist with technical terms.

1
Introducing Orchids

Without doubt the most fascinating group of plants on earth is the orchids. For centuries they have attracted attention and today they have an enthusiastic following in most countries of the world. Their beautiful flowers are widely used in corsages and for floral displays and in recent years flowering plants have become popular subjects for indoor decoration. The cultivation of orchids has been described as a disease because of the obsession with which growers pursue new plants for their collections. Hybridisation has long passed the stage of primary and secondary crosses with many growers having only hybrids in their collections. Recent estimates have numbered man-made orchid crosses in excess of 50 000, easily exceeding the number of species in the family and this could increase dramatically in the next decade. In many of these crosses the aim of "improvement" is to achieve fullness of flower and roundness of shape – a far cry from those flowers found in nature. A real danger is that future hybrids will all end up the same shape and much of the charm and diversity of the species will be lost forever.

DEFINING AN ORCHID

With the diversity exhibited by orchids this task is just about impossible. Some facts however can be assembled together on the subject. Orchids are herbs (having no woody tissue) and are placed in a major subclass of the plant kingdom known as the monocotyledons. Other plant groups found in this subclass are rushes, sedges, grasses, lilies, amaryllids, gingers, aroids and palms. All plants of this type have some features in common, in particular sheathing leaf bases, parallel venation and a single seed leaf.

Orchids are unique and differ from other monocotyledons by a number of singular features. The single seed leaf which characterises monocotyledons is often absent in orchid seedlings but is present in the reduced embryo of the seed. Orchids have three petals and three sepals which may be alike or dissimilar, however, the third petal is nearly always greatly modified and very different in size, shape, structure and often colour from the others. This is known as the labellum or lip, and it plays a major role in the pollination of orchid flowers. Labellums may be fixed, hinged on a claw so that they are mobile or they may even be actively irritable and capable of movement following a stimulus (*Caleana* spp.). They may be flamboyant, gaudy, waxy, slippery, glandular, hairy or have landing strips and guidelines marked out for their visitors. Often the labellum completely dominates the flower, rarely is it similar to the other petals (*Thelymitra*).

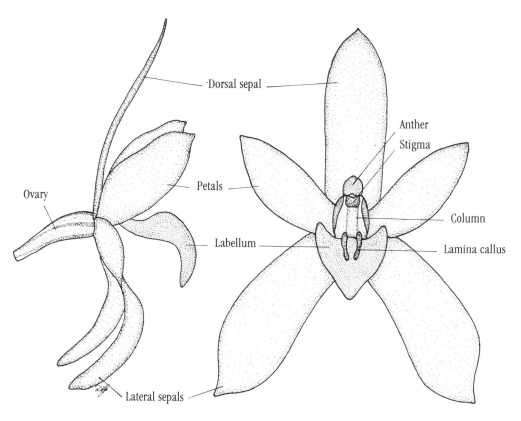

Fig. 1 Front and side view of *Cymbidium canaliculatum* flower showing main floral parts.

Another highly significant development of orchids is the combination of the male and female sexual parts into a central structure known as the column. This organ is often fleshy, may be quite conspicuous and sometimes has prominent staminodia which may be present as wings or ornate tufts of cilia. The anther and stigma are separate on the column (usually the anther is terminal), although often both organs are situated close together.

Orchid seeds are also highly interesting, being winged and with the embryo much reduced and lacking any significant differentiation into tissues. Orchid seeds are also minute and dust-like. Counts have shown that a single capsule may contain from about 1000 seeds to nearly 4 000 000, depending on the size of the fruit and the species.

THE ORCHID FAMILY

Botanists place orchids in the family Orchidaceae, a name introduced by the great English botanist and orchid specialist John Lindley in 1836. The name of the family and indeed the common name orchid used for the group in general, is based on the genus *Orchis* which was described by another great botanist, Carolus Linnaeus, in 1753.

Orchids are the largest and most successful plant group on earth. Estimates of their numbers range from 20,000 to more than 35,000 species in 750-850 genera. New species are being discovered regularly in the wild and others are recognised following detailed taxonomic studies by botanists. Orchids are a highly significant group of plants and their success can be gauged when it is realised that they comprise about 10% of the world's flowering plants.

GROWTH HABITS

Orchids can grow in the ground as terrestrials, perched on rocks as lithophytes or on trees as epiphytes. The term epiphyte is often used loosely to embrace both rock and tree dwellers since many orchid species commonly grow on both substrates.

Leaf-bearing terrestrials may be either evergreen with the leaves lasting for a number of seasons (for example, *Calanthe triplicata*, *Paphiopedilum* spp.) or deciduous with the plants dying back to a fleshy storage organ each year so as to avoid unsuitable climatic conditions. These storage organs may be situated in the top layer of soil (*Eulophia keithii*, *Calanthe vestita*) or are subterranean (*Habenaria* spp., *Orchis* spp.).

Leafless terrestrials are known as saprophytes. Such plants are devoid of chlorophyll and cannot produce their own compounds for growth and reproduction. They rely for their existence on a close relationship with a mycorrhizal fungus which invades the orchid's roots. Such a relationship is known as a symbiosis. Some remarkable saprophytes are strong climbers. Saprophytic orchids are often fleshy and brittle and may be pale-coloured, green,

Fig. 2 Two major growth forms of epiphytic orchids.

SYMPODIAL GROWTH

MONOPODIAL GROWTH

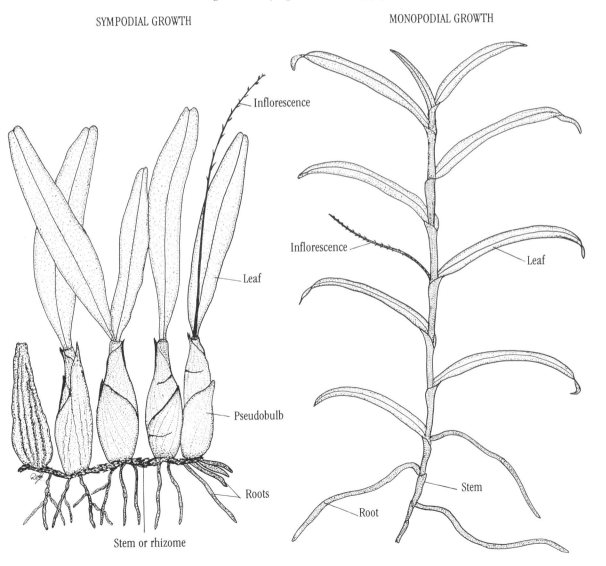

9

reddish or brownish. They also have irregular, fleshy root systems and as a general rule are impossible to maintain in cultivation.

The majority of orchids grow as epiphytes because the growth habit enables them to take advantage of niches in the forest canopy where suitable regimes of light and humidity exist. Most species of epiphytic orchids are evergreen however those growing in strongly seasonal climates may shed their leaves annually after the new growths mature. An interesting small group of epiphytes are leafless with their chlorophyll-bearing roots acting in the place of leaves.

Two significant growth habits within the epiphytes are easily identifiable and are of importance to orchid growers. Monopodial orchids have a main axis which increases in length steadily each year and the inflorescences and roots arise laterally from the side of each stem. Sympodial orchids grow in stages with each growth having a predetermined limit and maturing before a new growth is produced from its base. The growths are usually joined by a rhizome which may be very short. Roots are produced from the base of each growth and inflorescences arise from the top of the growth in some species and from the base in others.

DIVERSITY IN ORCHIDS

Much of the fascination of orchids arises from their great diversity, not only in growth features but also in flower form, shape and colour. In no other group of plants can one find such diversity as leafless epiphytes where flattened roots contain cholorplasts and take over the role of leaves (*Campylocentrum, Chiloschista, Microcoelia, Taeniophyllum*); lowly saprophytes that spend all their life underground (*Rhizanthella*); epiphytes ranging in size from those so tiny as to be comfortably grown in a thimble (*Notylia norae*), to giant clumps which may reach 7 m tall and weigh up to a tonne (*Grammatophyllum speciosum*); saprophytes which can generate so much energy that their climbing stems often reach 20 m in length and produce thousands of flowers (*Galeola kuhlii, Pseudovanilla foliata*); terrestrials which can only survive by an annual controlled invasion of a suitable fungus; epiphytes with male and female flowers so dissimilar that they have been described as different species (*Catasetum*). And so the list goes on.

The diversity in floral features of orchids is almost beyond comprehension. Inflorescences range from solitary flowers to spikes and racemes several metres long; they may be wiry, fleshy or flattened, erect, arching or pendulous and in some the flowers are borne in special, concave pits on the surface (*Bulbophyllum pachyrrhachis*). Flowers range in size from about 1 mm across in *Phreatia baileyana, Oberonia titania* to *Sobralia macrantha* which can reach 30 cm wide. They may last but a few hours (*Dendrobium crumenatum, Thrixspermum platystachys*) or weeks and even months (*Encyclia citrina, Cymbidium lowianum, Oncidium papilio*). The flowers of some may only open above a certain temperature (*Thelymitra*).

Perfumes are important for pollination and are an interesting additional aspect for growers. They are often very noticeable in the confined environs of glasshouses and can enhance the atmosphere. Some however are so strong as to be almost overpowering and leave growers with little doubt that they are in flower (*Stanhopea* spp., *Encyclia aromatica, Epidendrum fragrans*). Many orchids have a very pleasant fragrance which enhances the atmosphere (*Bifrenaria harrisoniae, Brassia caudata, Bulbophyllum suavissimum, Cymbidium traceyanum, Odontoglossum pulchellum, Zygopetalum mackaii*). Other attractive perfumes may be described as fruity (*Bulbophyllum fascinator, Dendrobium draconis,*

Fig. 3 Diversity in orchid floral shape

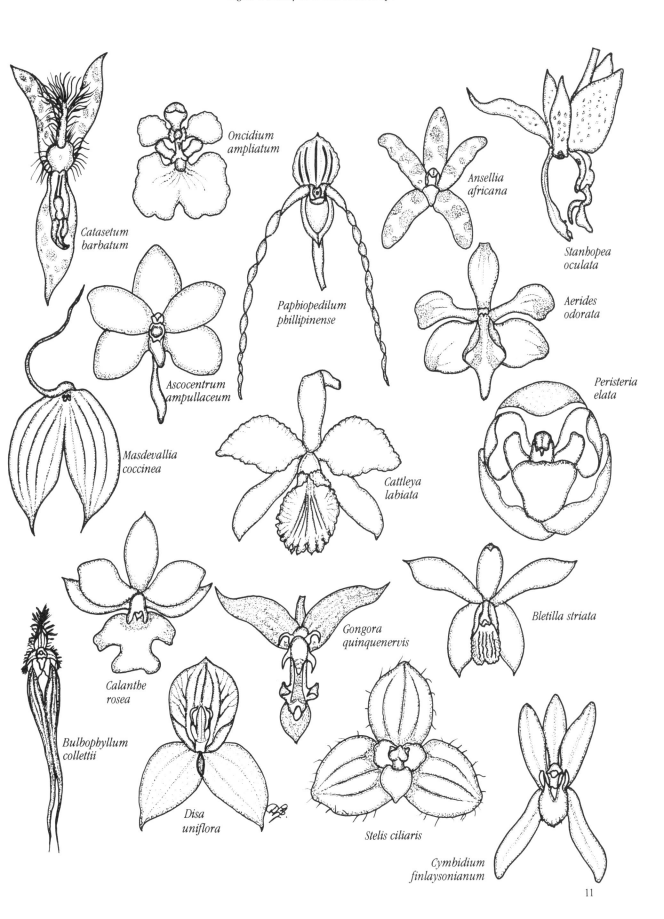

Catasetum
barbatum

Oncidium
ampliatum

Paphiopedilum
phillipinense

Ansellia
africana

Stanhopea
oculata

Aerides
odorata

Ascocentrum
ampullaceum

Masdevallia
coccinea

Cattleya
labiata

Peristeria
elata

Bulbophyllum
collettii

Calanthe
rosea

Gongora
quinquenervis

Bletilla striata

Disa
uniflora

Stelis ciliaris

Cymbidium
finlaysonianum

Lycaste candida, Maxillaria tenuifolia), spicy (*Gongora maculata, Lycaste aromatica*), reminiscent of honey (*Dendrobium luteocilium*) or even delicious (*Epidendrum phoeniceum*). Some perfumes are not particularly attractive being musty (*Bulbophyllum medusae, Coelogyne speciosa, Dendrobium moschatum*), sour (*Bulbophyllum crassipes, Dendrobium macrophyllum, Habenaria hymenophylla*) or downright foetid and obnoxious (*Bulbophyllum barbigerum, B. beccarii, B. dayanum, B. fletcherianum, Gongora stapeliiflora, Mormodes maculatum*).

WORLD DISTRIBUTION

Orchids are very widely distributed, being found on all the major continents of the world with the exception of Antarctica. Northwards they extend to Alaska and northern Sweden, both well within the Arctic Circle, while in the south they are found in Tierra del Fuega and Macquarie Island which borders Antarctica. Orchids have a tremendous altitudinal range which extends from sea level to nearly 4000 m, well above the tree line and almost to the snowline.

Orchids are most abundant and diverse in tropical regions and they become less common and less diverse with increasing distance from the equator. In temperate climates most orchids grow as terrestrials and epiphytes are rare whereas in the tropics the reverse is true with epiphytes predominating.

MYTHS, MAGIC AND SUPERSTITION

Many myths have surrounded orchids resulting in numerous misconceptions about them. For example it is often believed that those orchids which grow on trees rob their host of nutrients and act as parasites. This is incorrect for the orchid roots are limited to the outer surface of the bark (which is dead) and nutrients necessary for growth are picked up from here or in litter which collects in cracks and fissures. The sheer flamboyance, colouration and exquisite or even bizarre flowers that are commonplace in this group of plants have led to a mystical aura that is reinforced by the anthropomorphic interpretation of floral shapes. Thus we have lady orchid, man orchid, slipper orchid, monkey orchid, swan orchid, ghost orchid, dove orchid, scorpion orchid, donkey orchid, and more. Mystical beliefs, magic and superstition have long been associated with orchids and these have become entrenched in the folklore and religion of many cultures. Most of these relate to supposed aphrodisiac properties and love charms but other recorded uses were to ward off evil spirits, as death charms, as protective charms against enemies, to cure illness and in witchcraft. These beliefs are widespread and transcend religion, culture and race. For a detailed and fascinating account on this subject and others dealing with orchid use by man, see Ethnobotany of the Orchidaceae, by Len Lawler in Orchid Biology and Perspectives III, 1984, Cornell University Press, Ithaca and London.

ECONOMIC IMPORTANCE

Despite the very large numbers of orchids, only one product, vanilla, has achieved any economic importance. This material is extracted from partially ripe capsules (known as vanilla beans) of species of the genus *Vanilla* after a fermentation process. Vanilla is used in perfumes and as an aromatic food flavouring. Plantations for its commercial production have been established in Madagascar, Mexico, Central and South America, Africa, the West Indies, Tahiti and Fiji. The active principle of vanilla (vanillin) can be prepared synthetically but there is still economic demand for the natural product which is regarded as superior.

Orchid nurseries are well entrenched in many countries and the annual turnover of plants is probably in the millions. While there is a good trade in species, the real money spinners are the better hybrids especially award winners. These may be mass-produced by vegetative techniques such as meristem culture.

Cut flowers are also an important crop harvested from orchids. In South-east Asian countries hybrids having long spikes of long-lasting colourful flowers are harvested by the millions annually and exported to Europe, America and Australia where they are used for floral decoration. These orchids are grown out of doors or in lathe houses and revel in the tropical conditions. Seedlings are continuously being selected for new hybrids, tested and reproduced in large numbers if successful. Annual sales of these orchids are in the millions of dollars and the crop is a major revenue earner for the countries involved.

Other orchids may be used as cut flowers and significant industries are developed around hybrid *Cymbidium* in the USA and Australia and *Phalaenopsis* in the USA. *Cymbidium* flowers are used for decoration and as excellent corsages while *Phalaenopsis* are in great demand for weddings. *Cattleya* flowers are also popular for corsages but are mainly a local industry as they do not travel well.

ORCHIDS IN AUSTRALIA

Orchids have a very healthy following in Australia. Currently there are about 170 orchid societies and they are scattered throughout the country with very strong representations in the capital cities. Meetings are held monthly in most cases and there is always at least one flower show annually. These shows, predominately in spring, always attract considerable attention and are popular with the general public and orchid enthusiasts alike.

2

The Cultural Requirements of Orchids

Orchid plants are easy to grow as can be testified by the huge number of growers to be found in various countries around the world. It is easy for people living in the tropics to grow orchids but even in extremely cold climates it is possible, with ingenuity, to create suitable conditions and grow these appealing plants. We read in orchid periodicals of collections coddled in special glasshouses, converted rooms and even cellars and attics modified with heat and banks of artificial lights to provide conditions suitable for growth and flowering.

While it is easier to grow orchids in the tropics, it must be understood that not all orchids are suited to such conditions. Thus those species originating in temperate regions usually require some exposure to cold during the year and will languish or die if transferred to a tropical climate. Also photoperiods experienced in the tropics may be unsuitable for regulating growth and flowering. Thus it is difficult to grow temperate orchids in the tropics however the reverse is not usually true if heat is provided. A glasshouse, heated by modern technology, will avoid the extremes of cold and can provide suitable conditions for even the most tropical of plants.

A novice orchid grower usually is somewhat starry-eyed about the subject and tries to grow as wide a range of orchids as possible. Experience is the best teacher and hardened growers realize only too well the folly of trying to grow orchids which have vastly different cultural requirements together where conditions are similar. Thus dedicated growers end up with more than one glasshouse in their backyard and their plants are shuttled around until the optimum conditions are found for each. With such a tremendous range of orchids available to select from, it is not difficult to choose a group which will thrive in a particular set of conditions. The message then for novices is to gain as much knowledge as possible and to proceed steadily with an aim in mind rather than with reckless abandon. Knowledge can be gained from experienced growers, orchid periodicals and books. Joining a local orchid society is one of the best ways to start.

Understanding Orchid Requirements

The best orchid growers are those who have a good understanding of the requirements of their plants. One of the most effective ways to gain this knowledge is to observe orchid plants growing in their native state. Unfortunately this is not always possible and the next best approach is to read avidly. Orchid journals can provide a wealth of information not only in those articles directed at cultivation but also in the mouth-watering reports of collectors who have been fortunate enough to visit out-of-the way localities. Even a photograph of the terrain can provide clues. Note whether the vegetation is open or dense, the topography flat or steep, large boulders and cliff faces obvious or the mountain tops wreathed in mist. These are the sort of conditions that can influence the requirements of a species and an observant grower takes note of them.

The importance of gaining an understanding of the needs of orchids through a study of their natural habitat cannot be overstated. It can be exemplified by considering deciduous terrestrial species and also epiphytes from the tropics.

Deciduous terrestrial orchids are found in a wide range of environments. They are most prevalent in temperate regions but some species occur in the tropics. In general their dormancy period coincides with climatic extremes of heat and dryness, thus the dormancy is a mechanism to avoid stressful conditions. Those species from temperate regions normally begin growth in the autumn and flower in autumn, winter or mostly spring. Those species from very cold climates may have the need for winter chilling before they will flower in the increasing day length of spring or early summer. Contrast these temperate orchids with the tropical terrestrials where the arrival of the summer monsoons is the major stimulus. The first heavy rains induce a rapid response and the plants produce all their growth and flowers during the three or four months of rain. Thus it is obvious from these few notes that terrestrials from cold temperate climates would have slightly different cultural requirements from those in warmer temperate areas but both groups would have very different needs from those originating in the tropics.

To know that an epiphytic orchid originates in the tropics is very useful but precise information will enable a more accurate appreciation of its cultural requirements. Thus those species originating from highland tropical regions (where it can get very cold) have different needs from those from tropical lowlands where a temperature of 20°C can be considered chilly. Many highland species grow where clouds and mist gather and air movement of some sort is nearly constant. In the tropics rainfall may be spread throughout the year with a summer maximum or be strongly seasonal with a long dry period following the summer wet. Orchids from each will have very different needs. Variations exist even within a patch of forest. Thus those species which grow high on exposed trunks and branches dry out very rapidly after rain in contrast to those on mossy boulders and trunks where the moisture levels are much more constant.

TEMPERATURE

It is generally acknowledged that tropical orchids are sensitive to cold periods and are rapidly killed by frosts. It follows therefore that in unsuitable environments they must be grown in protected conditions such as are provided by heated glasshouses.

Orchids from lowland tropical areas are especially cold sensitive and may begin to suffer when temperatures fall to 8°C. Species with limited storage reserves, such as the vandoid orchids (e.g. *Phalaenopsis*), may collapse dramatically at low temperatures and have little capacity for recovery. Pseudobulbous types generally decline more slowly and have a better chance of recovery if moved to congenial conditions. In general lowland tropical species can be overwintered if maintained at a minimum temperature of 10°C, but there is a much better safety margin at about 15°C, especially if the winters are long and severe. Sensitive species may only be happy at 18°C to 20°C.

A very large range of orchids are much less cold sensitive than the above group but still may suffer if the temperature reaches 5°C. Short spells at this temperature may not be a problem but extended periods can be very detrimental. Orchids with these temperature requirements are often found in subtropical zones or at intermediate altitudes on ranges and tablelands.

Altitude considerably modifies tropical climates. Thus at levels of 1500 m and above, the summer temperatures are much less extreme than on the lowlands and the winters colder. As a general rule those species originating from high mountains in the tropics are quite tolerant of cold although in very severe climates they may need protection from frosts.

Overwintering Strategies

A few strategies can be employed to assist cultivated orchids to overwinter. Even those plants in a tropical glasshouse will suffer some retardation over winter; if not from the cold, then as a result of short photoperiods. Plants which are in active growth suffer much more from cold than those which have mature growths or are quiescent. Such plants must be coddled if the growth is to mature. Plants can be hardened by reducing watering in the weeks preceding winter and will then overwinter much more readily. Fertilizer application should be confined to spring and summer so that active growth is finished by the time winter sets in. Watering should be kept to a minimum during winter and the chilling effects of using cold water from outside pipes should not be forgotten. If possible have an internal water source or at least take the chill off it first. Cold sensitive species should be moved to the warmest part of the glasshouse prior to winter. Close to the glass near the ridge is often a good site.

HUMIDITY

Humidity is a reflection of the water vapour present in the air. It is also closely linked with temperature and thus at moderate temperatures high humidities are much less noticeable than they are at high temperatures.

Orchids are variable in their humidity requirements. Thus those species growing naturally on the floor of rainforests require almost constant high humidity whereas those from savannah woodlands are much less demanding. Some orchids may require lower humidity when flowering. It is generally difficult to maintain a constant high humidity and a level of about 50% is usually quite satisfactory for a general collection.

Humidity can be increased by watering, misting, the use of humidifiers and water-retentive materials such as sphagnum moss. High humidity may be good for plant growth but may also lead to disease problems. Wildly fluctuating humidity is not good because it may cause a stop-start situation and result in stunted growth.

Some orchids like to be dry themselves but need to be grown in a humid situation with adequate air movement. Many examples are to be found among the small vandoid orchids, especially those which grow in the drier, stunted types of rainforest. This type of environment is very difficult to achieve and is not found in most glasshouses. It can be approximated by suspending the plants above a large container of water or a pond within a glasshouse.

AIR MOVEMENT

Air movement is of major importance for the successful cultivation of orchids. It distributes oxygen, carbon dioxide and water vapour around the plants, prevents stagnant conditions and reduces the incidence of disease. Air movement should be gentle either being created by successful ventilation using changes in air temperature to induce an air flow through the glasshouse or by the use of internal fans. Excessive air movement may dry out plants and draughts are to be avoided at all costs.

LIGHT AND SHADE

Orchids grow in a wide range of habitats, some in nearly full shade, others in full sun and with all combinations in between. While it is difficult to be dogmatic about specific light requirements for cultivation some useful generalisations can be made.

Those orchids which grow naturally in full sun have high light requirements and generally perform poorly if grown in too much shade. They should be grown in the open or in an unshaded part of the glasshouse. Terrestrials from the rainforest floor have a much lower light tolerance and need to be protected from full sun. About 60-80% shade is usually satisfactory. In between these extremes orchids may grow in dappled light which filters through the forest canopy or in situations where they are exposed to full sun for short periods during the day. For these orchids about 30-50% shade is necessary.

It must be realised that in temperate areas orchids will only require shading during the summer months and for the rest of the year should receive good light. In the tropics however the shading may be needed throughout the year.

The hours of daylight received is known as the photoperiod. The length of the photoperiod (or more accurately the length of the dark period) is of major importance in stimulating orchid flowering. Long day species respond to decreasing periods of darkness (increasing daylength), others to the reverse while many tropical species are unresponsive to photoperiod.

WATER

Orchids in general have high water requirements especially during the warm to hot months of the year when they are in active growth. Paradoxically most species are also well adapted to withstand several days without water especially when growth is quiescent.

Watering is dealt with in more detail in the next chapter however some notes on water quality are pertinent here. Reticulated water may or may not be suitable for orchids depending on its quality and to some extent the species. *Disa* for example are rather sensitive to water quality and if the plants are not growing well this factor should be examined. If the water quality is poor, then rain water collected and stored in tanks is usually a suitable alternative. Hard water may cause problems especially that with high levels of calcium salts (particularly bicarbonates). Such water may cause nutritional imbalances in iron, magnesium and some trace elements which are very difficult to correct.

NUTRIENTS

Orchids are no different from other plants in that their roots extract nutrients from their surroundings. For terrestrial species this includes the soil, humus and decaying litter. For epiphytes the roots wander over rocks, the surface of bark and in crevices where detritus collects.

The elements required by orchids in order that growth, flowering and reproduction can occur are no different from those required by other plants. Some elements are required in large quantities and are known as major elements, while the minor or trace elements are needed in very small quantities. Both types are listed below.

Major Elements	*Minor Elements*
Nitrogen	Iron
Phosphorus	Manganese
Potassium	Boron
Magnesium	Zinc
Calcium	Copper
Sulphur	

Best growth is achieved if a balance of nutrients is available rather than if one or more are in excess. If a single element is deficient then normal growth is disrupted or distorted. Adding the missing element is the only way to restore normal growth. Increasing the levels of the other elements will bring no response until the missing element is restored.

Basic details of fertilisers and their use are outlined in the next chapter.

3
Epiphyte Cultivation

Methods of orchid cultivation vary in different countries and climates. In general it is much easier and cheaper to grow a wide range of orchids in the tropics than it is in temperate regions. In tropical regions growers often tend to use simpler methods of culture such as garden trees or stakes embedded in the ground.

Because aspects of cultivation used for epiphytes differ from terrestrials, both groups are dealt with separately. The bulk of this chapter details epiphyte cultivation and terrestrials are presented in the next chapter.

This chapter introduces the basic techniques needed to grow epiphytes successfully. Experience will put the flesh on these techniques.

EPIPHYTES IN POTS

The bulk of cultivated orchids are grown in pots. This is true for both epiphytes and terrestrials, irrespective of the climate. Pot culture has some advantages over other systems. Pots themselves are generally cheap, are easily moved about to find the best position for the plant and can be readily checked for dryness, pests and breakdown of the potting mix.

Suitable Containers

Traditionally orchid growers have used terracotta pots, usually with excellent results. Providing the drainage is adequate (enlarge the drainage holes if in doubt) and the potting mix suitable, then epiphytes grow extremely well in this type of container. The terracotta itself is a suitable substrate for orchid root growth with the roots clinging tenaciously, as anyone who has repotted orchids from these containers can testify. Terracotta pots are available with additional holes in the sides and these are excellent for epiphytes. Pots of a squat design are particularly suitable for orchids whose roots do not like to penetrate to the bottom of the potting mix. Terracotta pots have the disadvantage of being heavy and significant root damage is often caused at repotting time because of root adherence. This can be offset somewhat by pre-soaking in a tub of water and using a thin knife to lift the roots from the container.

Plastic pots are a relatively recent innovation but they have revolutionised the nursery industry and are now well entrenched among orchid growers. In general they are readily available, cheaper, lighter and maintain a more uniform water regime than do terracotta pots. Squat designs are often better than deep pots because they drain more freely. Often the drainage holes are masked by flaps of waste plastic and these should be removed before use. Drainage holes are easily enlarged by the use of a heated knife blade.

Some growers prefer plastic pots, others swear by terracotta. Each can be very successful in its own way and it really comes down to availability, grower preference and local conditions. Problems are certainly created with water regimes when orchids in terracotta and plastic pots are mixed together. Small terracotta pots are also much better than plastic pots for those tiny botanical treasures which really suffer if overpotted.

REQUIREMENTS OF AN EPIPHYTE MIX

Epiphytic orchids need excellent drainage and plenty of aeration around their roots. For this reason they cannot be grown in soil and must be potted into a coarse mix that is free of soil and other fine, clogging particles.

Suitable Potting Materials

Softwood Barks: These materials, popular with orchid growers, are obtained as a waste product from the softwood timber industry and are available in many countries. Bark suitable for orchid growth can be obtained from pines, redwood and firs. The best bark, which is nuggety or flaky, is obtained from large, mature trees. That from slender, immature trees is difficult to break, remains fibrous and is often contaminated by wood slivers. Fresh softwood bark may contain toxic materials which can inhibit orchid root growth. If the sample has a strong, resinous smell it is best treated prior to use. Treatment consists of saturating the whole sample with water and storing it in a heap for about six weeks, keeping it moist all the time. Suitable softwood barks have a pH of 6-6.5. The pH of pine bark is initially about 5 but rises with ageing to 6.5.

Charcoal: This material is widely used for the cultivation of orchids. It can be sieved to produce different grades and should be washed prior to use to remove dust. It is usually mixed with other materials but some growers use it exclusively. Its qualities vary somewhat depending on the wood from which it is produced. Charcoal can absorb nutrients and salts may build up to a toxic level unless it is occasionally heavily flushed with water.

Osmunda Fibre: This black, wiry fibre is actually the root system of the fern genus *Osmunda*. It is obtained from large clumps of the fern and has been used traditionally to grow orchids for generations. It can be used very successfully as a potting material by itself and in fact it does not mix well with other materials unless chopped. Osmunda fibre contains most nutrients which orchids need and it produces excellent growth.

Treefern Fibre: This material consists of the chopped up fibres which make up the trunk of a treefern. It is one of the best materials which can be used in an epiphyte mix as the long fibres keep the mix open ensuring drainage and aeration while at the same time having some water-holding capacity. With hundreds of treefern species scattered around the world it can be expected that some will be more suitable than others. Material obtained from long-dead trunks may be acid or require pre-soaking in a solution of lime prior to use. Some growers use treefern fibre in chunks or cubes as a component of potting mixes.

Polypodium Fibre and Staghorn Peat: These peaty, fibrous materials are obtained from large fern clumps usually after they have died. Such materials are popular with orchid growers and can be used successfully for some species. They tend to hold a lot of water and at the end of their useful life, usually deteriorate very quickly.

Sphagnum Moss: This material is commonly used as a potting mix by some growers for small epiphytes which have very fine roots. It is also excellent for species of *Disa*. Only good quality, light-coloured moss is used and generally in small containers. Chopped sphagnum moss is sometimes added to a potting mix to improve its water-holding capacity.

Coarse Sands, Gravels and Grit: These materials are added to epiphyte mixes as a safeguard, for if the potting mix should break down, such coarse materials will ensure some drainage. All of these materials are heavy and some growers may prefer to use perlite or polystyrene to produce similar results.

Blue Metal, Scoria and Stone Chips: These materials have a similar use to coarse sand for keeping a mix open. In some local areas, however, growers may use them exclusively as a potting material. Particle sizes about 6-10 mm across seem satisfactory. Limestone chips may be added to the potting mixes of orchids which grow naturally on limestone e.g. some *Paphiopedilum* spp.

Terracotta and Brick Shards: Broken pieces of terracotta and brick can be useful when added to a mix, especially for large orchids. They provide excellent drainage and it is also noticeable that the roots often adhere to them.

Peanut Shells: These can be added successfully to mixes used for epiphytic orchids. Coarse, unchopped shells are best. They breakdown after a couple of years and should not be used at more than 10% of a mix.

Leaf Mould: Some coarse, partly decomposed leaf mould can be usefully added to an epiphyte mix. About 10% is sufficient.

Suitable Potting Mixes

A whole range of potting mixes suitable for epiphytic orchids can be prepared from the above materials and indeed others. There are nearly as many mixes as there are orchids. Growers are often influenced by local conditions and should be continually experimenting and observing the resultant growth of their orchids.

Some materials such as osmunda, softwood bark and charcoal can be suitable by themselves or may benefit from the addition of other materials. Cost may be a consideration, especially in large collections, and softwood barks may be expensive compared with charcoal. Mixes for hanging pots need to be light whereas on the benches the addition of some heavy material will ensure stability.

As a guide only, a useful basic mix can be prepared from 4 parts bark, 2 parts charcoal and 1/2 part peanut shells or treefern fibre.

Particle Size

Small-growing epiphytic orchids, in particular those which have very fine, slender roots, respond best when potted into mixtures of fairly small grades e.g. particle sizes 5-10 mm across. Large-growing or vigorous epiphytes with coarse roots generally require a lot of aeration in their mix and are much better when grown in coarse, chunky mixtures e.g. particle sizes 20-40 mm across.

To obtain some uniformity the main potting materials should be seived into fine, medium and coarse grades prior to mixing.

POTTING AND REPOTTING

It becomes necessary to repot epiphytic orchids when
1) the potting mixture begins to break down and impedes drainage and reduces aeration. A useful guide to deteriorating pottings mixes is given by the presence of moss or algae on the surface or of persistent weeds.
2) the pot is so full of roots and the mixture exhausted of nutrients that growth is suffering.
3) when the plant outgrows its container (see also under Potting On)
4) the mix is unsuitable for local conditions. This can apply to plants obtained from other growers or purchased from nurseries.

Repotting is best carried out over the summer months with an optimum time just before the new growths are produced. Some growers prefer to delay until the new growths have started but then there is always a danger of damage. Repotting just prior to the appearance of new growth has the distinct advantage that new roots are produced rapidly from the base of the growths and these quickly establish in the fresh potting mix. Winter is not a good time for repotting unless it is the only way that an orchid's life be saved.

Plants are prepared for repotting by removing all the old potting mix, dead roots, leaves and pseudobulbs. The potting mix can usually be shaken free but some may adhere to the roots. The plant should be examined for pests and any present are eradicated before repotting.

The procedure for potting and repotting orchids is identical and is outlined in the next section.

Potting Technique

Growers vary in their potting techniques but the basic moves are similar. Hold the plant in place in the pot so that its base is nearly at rim level and fill with potting mix. Pack the potting mix firmly around the roots so that the plant is supported. Tapping the base of the pot on a bench is a good way to settle the mix. The potting mix should be level with the base of the plant. Rhizomes and any new growth should not be buried. Some growers fill the pots to the rim, others leave a catchment of 2 – 3 cm to aid watering. Newly potted plants should be labelled immediately and placed in their growing area as soon as possible.

Plants with a poor root system may require staking to prevent excessive movement until new roots are produced. If the potting mix is too dry it should be moistened before use. Some materials can become excessively dry and are then difficult to wet especially in a pot. Water just runs down the side without actually wetting the mix and the plants suffer from dryness.

Potting On

Potting on is the simple procedure of tipping out an established orchid and potting into a larger container without greatly disturbing the root system or potting mix. This technique should only be employed for strong-growing orchids in which the potting mix has not begun to deteriorate. As the plants are repotted into successively larger pots, a coarser grade of mix is used. For sympodial orchids, the plant should be placed into the new pot so as to give maximum growing space before it reaches the edge. Any gaps are filled with new potting mix, the plants watered and placed back into the collection.

WATERING

The watering needs of epiphytic orchids vary depending on such factors as the season, temperature, light intensity, degree of air movement and the health of the plants. In the tropics, orchid plants may be watered two or three times a day and the same may apply in the summer months to plants grown in glasshouses in temperate regions. In the winter however, plants need much less water and in temperate climates may only be watered every seven to ten days, even in glasshouses. In the tropics in winter, intervals of three to five days may be acceptable. Plants grown outdoors generally dry out more rapidly than those housed in the confines of a glasshouse. It should be remembered that orchids can withstand short periods of dryness and more orchids die from overwatering rather than underwatering.

Watering should not be confused with misting or damping down which is carried out to enhance humidity. Watering consists of thoroughly drenching the potting mix and root system until excess water flows out the drainage holes. Some growers prefer to water early in the morning, others in the evening. The time of watering is probably unimportant in the summer months but in winter it is best to water in the morning so that plants have all day to dry out.

FERTILISERS

Epiphytic orchids benefit from the use of fertilisers and usually respond by producing strong healthy growth, large firm leaves and spectacular flowering. The frequency of fertilising depends to a large degree on the climate and the type of orchids being grown. In tropical regions orchid plants may be fertilised throughout the year however in temperate areas fertilisers are only applied during spring and early summer when the plants have a long growing period ahead. In these climates late applications of fertilisers may delay dormancy and interfere with flowering.

Strong-growing orchids have higher nutrient requirements than slow growers or miniatures and should be fertilised more frequently. In South-east Asia vigorous orchids which are grown for cut flowers may be fertilised daily with liquid solutions. Animal manures are also added regularly to the beds containing their root systems.

Type of Fertiliser

Organic fertilisers are excellent for orchids because they release their nutrients in a slow, sustained manner over a period of time. Blood and bone, hoof and horn and bone meal are fairly commonly used to promote healthy growth. Weathered, dry animal manures can also be equally successful. Slow release fertilisers are a modern alternative which some growers have taken up with alacrity. These include compressed pills and polymer-coated compounds with a release rate related to temperature. Fertilisers of these types can be incorporated in the potting mix or added to the top of established plants.

Liquid fertilisers are easy to apply and are an excellent means of maintaining healthy growth. Many commercial mixtures are available and some growers prefer to concoct their own or modify a commercial one to their needs. Organic extracts such as fish emulsion and seaweed extracts are very popular with orchid growers and produce excellent growth. A drawback with fish emulsion is its strong, pervading smell.

PESTS AND DISEASES

Pests and diseases which cause problems with cultivated orchids are dealt with in Chapter 5.

EPIPHYTES IN BASKETS

Many species of epiphytic orchids adapt very well to baskets. Vigorous vandoid orchids and sympodial species with long-creeping rhizomes grow especially well in this type of container. Species of *Stanhopea* produce strongly pendulous flower spikes and must be grown in this type of container if they are to flower properly.

Hanging baskets offer some advantages over pots. Drainage and air movement are enhanced as some orchids like to have their roots growing through the air. The baskets can be suspended near the glass or below benches for shade-loving species.

Types of Baskets

Three types of baskets are suitable for the cultivation of epiphytic orchids. Plastic baskets are generally unsuitable. Terracotta and plastic pots can be suspended successfully.

Slat Baskets: These are handmade from slats of wood and are excellent for many types of orchids. The timber should be hardwood, preferably redwood or teak which are resistant to rotting. In South-east Asia strong-growing vandoid orchids are commonly grown in empty teak baskets. These give maximum aeration and total drainage around the roots for these orchids in which the roots often like to grow through the air. For other species the basket is filled with a suitable potting mix after the gaps between the slats have been lined with a material such as sphagnum moss or coconut fibre.

Treefern Baskets: These are made from hollowed sections of dead treefern trunks. They have the advantages of free drainage combined with good water-holding capacity and the orchid roots can penetrate between the fibres. Disadvantages are that treefern baskets may have a limited life especially if kept continually wet and they may become more acidic with age.

Wire Baskets: These containers are commonly used for a wide range of plants and can be suitable for epiphytic orchids. The wire should be galvanised or else covered with plastic to prevent corrosion. Wire baskets must be lined with some material to retain the potting mix. Suitable materials include sphagnum moss, staghorn peat, paperbark and coconut fibre. Pre-cut fibrous basket liners are also available commercially.

Potting Mixes

Potting mixes used for hanging baskets are generally the same as those for pots. As a general rule the potting mix must be very well drained, however some growers add chopped sphagnum moss to enhance the water-holding capacity of the mix. When planting baskets the mix should be firmed down to leave a catchment of 2–3 cm.

Watering Baskets

Hanging baskets generally dry out rapidly because of the continuous circulation of air around them. This is enhanced in windy weather especially if the baskets are hung in shade houses or in windy sites. Baskets then must be watered more frequently than pots.

Fertilisers

Orchids in baskets benefit from fertiliser applications (see section under Orchids in Pots).

EPIPHYTES ON SLABS

Slab culture is a well-entrenched cultivation technique that has been used successfully for epiphytic orchids for many decades. Many species seem to prefer growing on slabs because of the extra air movement and the rapid drying which occurs after watering. Many also seem to like having their roots exposed to the atmosphere rather than being buried in potting mix as occurs in other systems of culture. Those species with a creeping habit or pendulous stems grow best on a slab whereas plants with crowded, erect growths are better accomodated in a pot. Plants grown on a slab are easily moved and can be shifted about until a suitable position is found. Orchids on slabs require fairly high humidity and bright light.

Slabs can be made from various materials. Some of these have a reasonable water-holding capacity whereas others dry out rapidly after watering.

Suitable Types of Slabs

Treefern: There are many hundreds of species of treefern scattered around the world and the trunks provide material which is used for orchid cultivation in many countries. Slabs of treefern are fibrous, drain freely and are thus well aerated while still retaining some water between the fibres. They vary greatly in consistency depending on the species. Some have coarse fibres, others are fine and wiry and yet others may have numerous root hairs present; some have parallel fibres while in others these form an interlaced network. Old trunks may have become acid with age and soaking in limewater for one to four weeks can be beneficial. Pests such as slugs, earwigs, slaters and cockroaches may live and breed between the fibres and emerge, usually at night, to feed on the orchid roots.

Weathered Hardwood: Slabs of hardwood, which have been weathered by exposure to the elements, are excellent for the cultivation of a range of epiphytic orchids. Such slabs do not absorb much water and dry out rapidly after watering and hence are suitable only for humid conditions and hardy species.

Natural Cork: This material is in fact the outer bark of the Cork Oak (*Quercus suber*). It is excellent for the cultivation of some species but others do not take to it.

Compressed Cork: This material consists of pieces of cork combined with glue under pressure to form a combination sheet. It is used successfully by some growers but often the orchid roots tend to avoid the surface rather than clinging tenaciously.

Pieces of Branch: Sections of tree branches can be used to good effect for some orchids but not all trees are suitable. One of the best is English Oak (*Quercus robur*) because it retains its bark for many years and orchids grow strongly on it. European Plum (*Prunus domestica*) is also useful as are many species of *Casuarina*. Paperbark species of *Melaleuca* are often used in the tropics but can be short lasting and often suffer from borer attacks. Coconut husks are another material often used in the tropics.

Tying

Orchids should be attached to slabs and tied tightly so they cannot move. Suitable materials include copper wire, nylon fishing line or strips of nylon fabric (panty hose or stockings). Materials which stretch and loosen after a short time are unsuitable.

Watering

Slabs dry out more quickly than pots and hence require more frequent watering. In the summer two to four waterings a day can be beneficial. In the winter plants may only need watering every second or third day.

Fertilisers

Orchids grown on slabs benefit from fertiliser application (see section under Orchids in Pots).

EPIPHYTES ON TREES

In areas which have a suitable climate, epiphytic orchids can be grown on garden trees. This is especially true in the tropics where a wide range of species can be cultivated by this simple technique. Prevailing climatic conditions are ideal and if the plants are watered in dry spells and given occasional fertilising, the results can be spectacular. Success can also be gained in the subtropics but here the range of species which can be grown by this method is somewhat less. Even in temperate regions a number of cold-hardy epiphytes take well to this type of culture.

It is an interesting observation that some species which can be difficult to grow well on slabs or in pots grow strongly when attached to living trees. Some species prefer a degree of protection and perform best on the sheltered side of a trunk whereas others require bright light. Some like plenty of moisture and others prefer to be perched where their roots dry out rapidly after watering. Large plants may need to be supported and siting them in the crotch of a tree may be an advantage. Observation of the trees during wet weather will show where the water runs off the trunk and branches.

Suitable Trees

Not all trees are suitable for the culture of epiphytes. Those which shed their bark are obviously unsatisfactory and some may have chemical toxins or inhibitors present which are antagonistic to epiphytes. Trial and error is the best guide. Many commonly grown garden trees are quite suitable for epiphytes, some more so than others. In the tropics calabash, frangipani and various species of citrus are very popular. In temperate regions species of ash, sheoke and deciduous fruit trees (especially *Prunus* spp.) can be successful.

Tying

Epiphytic orchids can be attached to trees by a range of materials including nylon fishing line, copper wire, plastic-coated wire and strips of stretch fabrics such as nylon stockings or panty hose. The latter materials are excellent because they expand as the tree grows. Wire should be monitored for damage to the tree and removed immediately it begins to cut into the bark. Once the orchid plants become attached by their own roots they are generally self supporting.

Watering

Initially the orchid plants will require regular watering until their roots become established. Established plants can withstand quite dry conditions but supplementary watering in very hot weather or long dry spells is usually beneficial.

Fertilisers

Occasional applications of liquid fertiliser or powdered animal manure promote strong growth.

Pests

The establishment of orchid plants on trees can be hindered by the activities of slaters, earwigs and millipedes. These feed actively on growing root tips (for more details see Chapter 5).

STAKES, STUMPS, ROCKS

Orchids are very adaptable, especially the hardier ones, and some unusual systems can be employed for their culture.

In the tropics, orchids are commonly grown on hardwood stakes or plant trunks which are embedded in the ground specifically for that purpose. These are often seen decorating gardens. Tree trunks or large branches may be used as well as cycads, treeferns and palms.

Tree trunks can also be successfully employed for the growth of epiphytic orchids. Hard narrow stumps are excellent but even hollowed stumps rotting in the centre can be used for orchids such as *Cymbidium*.

Boulders and piles of rocks can be used to support orchids, especially larger-growing species which grow naturally as lithophytes.

4

Terrestrial Cultivation

Terrestrial orchids are dealt with separately in this chapter because in general they have different requirements from those of epiphytes. Even within the large assemblage of orchids that grow terrestrially, variations in growth habit can indicate different cultural requirements. Those species found in rainforests are usually evergreen and this feature may also be prominent in terrestrials that grow in marshes and swamps. By contrast, terrestrials originating in regions which have extremes of cold or dryness are usually deciduous for part of their life cycle. These have tuberoids which may be buried in the soil or lie on the surface where they become covered by litter. Whereas the adaptable types of terrestrial orchids can be grown in mixed collections others are more difficult and require specialised treatment.

EVERGREEN TERRESTRIAL ORCHIDS

For convenience the shade-loving types, found in rainforests and other shady forests, are discussed separately from those which grow in sun and bright light in wet areas such as swamps and marshes.

Shade-lovers: Terrestrials of this type often have large leaves which are commonly thin-textured and pleated. They are easily burnt by excessive sun or bleached in too much bright light. Best leaf colour is achieved in about 70% shade and plants should be kept moist throughout the year. Although they require high humidity and shade, orchids of this type resent stagnant conditions with some air movement being essential. Repotting becomes necessary every two or three years although some vigorous species may require annual attention. Suitable potting mixes are based on such materials as chopped fern fibre, chopped sphagnum moss, coarse grit and leaf mould. Generic examples: *Acanthephippium, Calanthe, Corymborkis, Eulophidium, Paphiopedilum, Phaius, Tainia, Tropidia.*

The jewel orchids are an interesting group of terrestrials that are popular with some growers. They lack pseudobulbs but have fleshy, irregular stems and very thin-textured leaves. In general they require dense shade (70-90%) and constant high humidity. Some growers cultivate these in specially partitioned sections of a glasshouse or in structures such as bell jars, Wardian cases and large fish tanks. Potting mixes for these orchids are made up from chopped fern fibres, chopped sphagnum moss, fine seived pine bark, coarse grit and leaf mould. Generic examples: *Anoectochilus, Cheirostylis, Goodyera, Hetaeria, Ludisia, Macodes, Vrydagzynea, Zeuxine.*

Sun Lovers: Terrestrials of this group commonly grow in sunny situations usually where the soils are moist to wet over much of the year. In cultivation they require bright light, high humidity (70-90%), ample air movement and plenty of water. Suitable materials for a potting mix include well-drained loam, chopped fern fibre, coarse grit and leaf mould. Generic examples: *Arundina, Bromheadia, Neobenthamia, Spiranthes*.

DECIDUOUS TERRESTRIAL ORCHIDS

For convenience these orchids can be divided into two groups. The first group has corm-like storage organs (buried or on the soil surface) which have obvious nodes. The second group have tuberoids. These are swollen storage organs formed from roots and lacking nodes. The cultivation requirements of each group are different.

Cormous Group: A wide range of terrestrial orchids share storage organs of this type and many species are commonly grown. In some the storage organs are aggregated at the soil surface, whereas in others they are buried. They have a strict cycle consisting of a dormant period followed by active growth, flowering and seed production before dying back and entering dormancy again. When plants are dormant they should be kept on the dry side and repotted at the first signs of new growth. While in active growth the plants should be kept moist and warm. Potting materials can be prepared from such materials as chopped fern fibres, fine pine bark, good quality loam, chopped sphagnum moss, leaf mould and coarse grit. Those species which have their storage organs situated at the soil surface should not be buried too deeply. Generic examples: *Bletia, Bletilla, Calanthe, Geodorum, Malaxis, Pleione, Spathoglottis, Thunia*. Species with subterranean storage organs should be buried 5-6 cm below the soil surface. Generic examples: *Arethusa, Bartholina, Calopogon, Cypripedium, Disa, Eulophia, Pachystoma, Nervilia*.

Tuberoid Group: Orchids having this type of storage organ are often regarded as being very difficult to grow because they rely largely on an annual infection by a suitable mycorrhizal fungus for their survival. Because they have fairly specific cultural requirements they are not suitable for mixing in a general collection. Specialist growers, however, can achieve considerable success with orchids of this type. A major requirement is the inclusion of wood shavings in the potting mix as this provides a suitable substrate for the mycorrhizal fungus. These orchids are also very sensitive to root-rotting fungi and drainage of the potting mix must be excellent. Repotting should be carried out every one or two years. Considerable research has been carried out into the cultural techniques of these types of orchids in Australia and a successful potting mix is as follows: one part good quality loam, three parts coarse grit, one part leaf mould, two parts shavings. Generic examples: *Aceras, Acianthus, Caladenia, Calypso, Corybas, Diuris, Habenaria, Himantoglossum, Listera, Ophrys, Orchis, Peristylus, Pterostylis*.

Type of Pot

As a general rule, terrestrial orchids are grown in pots and are not suited to cultivation in baskets or on slabs.

Plastic or terracotta pots can be suitable for the cultivation of terrestrial orchids. Plastic pots are lighter, cheaper and maintain a more uniform soil water regime. Squat designs are particularly useful for terrestrials. Attention must be paid to the drainage holes to ensure that they are sufficiently large and unimpeded.

Potting Materials

Terrestrial orchids require a much closer potting mix than do epiphytes. Drainage must be free and unimpeded but the particle size of the materials used is much smaller than for epiphytes. Soil is not essential in a terrestrial mix however good quality loam is certainly beneficial.

Materials which may be used in preparing mixes for terrestrial orchids are listed below. Further details on some of these may be found in chapter 3.

Loam: Good quality loam has a pleasing friable feel or obvious fibrous texture which an experienced grower can recognise readily. The most suitable loams are obtained from natural forests where weeds are not a problem and a layer of litter and humus is present. A good quality loam has a stable structure which is maintained when used in a potting mix. Poor quality soils are not a substitute and are best avoided. Their structure can break down resulting in the drainage pores becoming clogged and such soils often set hard on drying. A simple test is to place a sample of soil in a small pot, flood it with water, then observe its rate of drainage and its consistency after a couple of days.

Chopped Fern Fibres: The chopped fibres of *Osmunda* and treeferns are an excellent ingredient for terrestrial orchids. They are of a suitable pH, excellent drainage and have good water holding properties.

Chopped Sphagnum Moss: This material is added to a potting mix to improve its water-holding capacity. It is also an excellent potting material by itself for species of *Disa*. Some growers cover the surface of the potting mix with chopped sphagnum moss to help keep the mix uniformly moist.

Leaf Mould: Leaves which are in a crumbly, partially-rotted state are an excellent additive to terrestrial mixes. When squeezed, a sample of suitable leaf mould should be crumbly and yet it retains a spongy, fibrous texture. Fresh leaves and undecomposed leaves are unsuitable.

Softwood Bark: Finely ground samples of softwood bark (particle size less than 0.5 cm) can be a useful additive to terrestrial mixes.

Wood Shavings and Buzzer Chips: These are included as a source of carbon for mycorrhizal fungi. They can be either hardwood or softwood but sawdust is unsuitable. Buzzer chips from Western Red Cedar, pressure-treated pine and chipboard mills should not be used.

Coarse Sands, Gravels and Grits: These materials drain freely with good aeration but they have no water-holding properties. Samples with angular grains are more efficient than those with rounded grains which tend to pack. Contamination with silt, clay and weed seeds may be a problem and a thorough washing before use is recommended.

Suitable Potting Mixes

Suitable potting mixes for terrestrial orchids can be prepared by mixing the above components in the following proportions.

	Mix 1	Mix 2
loam	1 part	
sand	3 parts	1 part
leaf mould	1 part	1 part
fern fibre	1 part	1 part
sphagnum moss		1 part

Variations can be made to suit local conditions or different orchid species by increasing or decreasing the proportions of any material or by adding other components such as softwood bark. Wood shavings and buzzer chips are added to mixes used for species which require the presence of suitable mycorrhizal fungi. Always remember as a basic rule that the drainage of the mixture must be excellent.

Watering

Evergreen terrestrial orchids should be watered in a similar manner to epiphytes, that is heavily during the spring and summer, less in the autumn and sparingly over winter.

Deciduous terrestrial orchids should be watered strictly according to their growth cycle. When dormant, watering should be limited to a light sprinkle every ten to fourteen days so as to prevent the mix from drying out excessively. Excessive watering during dormancy can lead to rotting.

Regular watering commences when the new shoots appear above the soil surface. Sufficient water should be applied to fill the catchment of the pot to the rim and excess will run out the drainage holes. The frequency of watering will depend on the prevailing conditions, but while the plants are in active growth the potting mix should be kept evenly moist.

As growth and flowering finish and the plants begin to enter dormancy the frequency of watering is reduced and the potting mix is gradually allowed to dry out.

Fertilisers

Terrestrial orchids in general respond to the use of fertilisers in much the same way as do epiphytes (see warning below concerning some deciduous types). Organic materials such as blood and bone, hoof and horn and bone meal can be beneficially added to potting mixes and slow release fertilisers can also be effective. Regular applications of liquid fertilisers over the summer months are an excellent means of promoting healthy growth.

Those terrestrial orchids which have a strong mycorrhizal relationship should be fertilised sparingly lest the balance be disturbed. Blood and bone, bone meal and hoof and horn can be added to the mix (10 g per 9 litres of mix) and liquid fertilisers reduced to about half strength.

Repotting

Terrestrial orchids are usually repotted every one or two years. Species with a strong root system or those which increase in numbers prolifically will require more frequent attention than the slow-growing types. Inspection of the potting mix will reveal whether the drainage pores have become clogged. The presence of moss or weeds in the pot are useful indicators that the mix is deteriorating.

Time to Repot

Evergreen terrestrial orchids are best repotted in the spring or early summer before active growth begins. Deciduous terrestrial orchids should be repotted while they are dormant as they are easy to handle then and damage is kept to a minimum.

5

Pests
and Diseases

While there are some pests and diseases which are a regular source of concern to orchid growers, overall this group of plants does not suffer as badly as many others. It is true that strongly growing, healthy plants are better able to withstand attacks by pests and diseases and in fact probably suffer less from them. General cleanliness and hygiene in the growing area is of major importance as is also the provision of fresh air by adequate ventilation.

A plant may be continually attacked by pests because there is something wrong with its health; perhaps it needs repotting or maybe it is potted into an unsuitable mix, is deficient in nutrients or does not appreciate its surroundings and needs to be moved.

If a pest or disease is present it should be identified accurately before control measures are taken. This may entail taking samples to an entomologist or plant pathologist at the nearest office of the Department of Agriculture. Accurate identification is of paramount importance in determining the control system to be adopted. Act as soon as the problem is noticed because pests and diseases can build up very quickly. If action is delayed then control can be much more difficult to achieve.

Pest and disease control in orchids usually requires the application of chemical sprays. This must be carried out with due consideration of the environment and all care to the sprayer. Protective clothing must be worn when handling any chemical and this must include a face mask or respirator. In the confines and generally high humidity of a glasshouse, the toxic effects of chemicals are reinforced. Once spraying is complete the area should be vacated, all doors and vents shut and visitations stopped until it is safe to do so.

Chemical sprays should be used sparingly and only if absolutely necessary. Some growers spray regularly whether pests are present or not. This practice should be discouraged because it selects resistant strains of pests and if these build up they become increasingly difficult to control. Also it is not good for the health of the grower to be regularly contacting plants covered with pesticide. Some chemicals may cause spray damage to foliage and buds, and should be used sparingly.

For convenience the major pests encountered by orchid growers have been grouped together loosely in this chapter according to their system of feeding.

CHEWING PESTS

The following pests have chewing mouthparts. Their activities are usually advertised by chunks being eaten from buds and leaves or grazed areas on the surface of organs.

Slugs and Snails

Slugs and snails are among the most destructive and persistent pests of orchids. The warm, humid conditions favourable for the growth of orchids are also conducive to these pests. As a group, they are very fond of the succulent tissues of orchids and frequently cause considerable damage to new shoots, flower buds and root tips. They feed by grazing and where the surface tissue is eaten, the damaged areas are at first pale green and then become papery and die. Damaged areas exude sap, become slimy and can provide an entry point for diseases.

Slugs and snails are mainly nocturnal and while they become most active during periods of rainy weather, their feeding is by no means curtailed by dry periods. The native species of slugs and snails are harmless but many of the twenty or so introduced species are highly destructive. The Common Garden Slug (*Deroceras caruanae*) is a significant widespread pest. Adults are 20-35 mm long, slender, grey-brown with a lighter coloured saddle. The Common Garden Snail (*Helix aspersa*) is a familiar sight to most orchid growers. Adults have a large, hard, yellowish brown, mottled shell and the eyes are carried on the end of sensitive, retractile stalks. Smaller snails commonly found in glasshouses are the Bradybaena Snail (*Bradybaena similaris*), which has a dark, spiral band on the shell, the Garlic Snail (*Oxychilus alliarius*), which smells strongly of garlic when squashed, and the Tiny Spiral Snail (*Cochlicella barbara*) which has a conical, whitish shell about 10 mm long. These smaller snails can hide in the potting mix and are destructive of seedlings and orchid roots.

Control of slugs and snails is usually achieved by baiting with pelleted commercial preparations of metaldehyde or methiocarb. Baiting in rainy weather is more effective than at other times. Pellets should be applied carefully, especially in glasshouses where they can become mouldy quickly in the humid atmosphere. This mould can spread into sensitive plant tissues, causing damage. Because of this problem some growers prefer to spray with liquid preparations of snail bait. Searching at night with the aid of a torch is a useful means of keeping the numbers of slugs and snails in check. Developing flower spikes can be protected from slugs by a wad of cotton wool wrapped around the stem near the base. If this wad is kept dry then slugs will not cross. Stale beer is a strong attractant and can be used in a simple drowning trap. Garlic snails are not attracted to baits but will congregate on the undersurface of a piece of fresh apple. If these small snails are a problem, thin slices of fresh apple strategically placed in the glasshouse each evening and collected the next morning can be an effective means of reducing their numbers.

Orchid Beetle or Dendrobium Beetle (*Stethopachys formosa*)

Various species of orchid beetles are to be found in the tropical regions of South-east Asia and Melanesia. They are very destructive pests of epiphytic orchids and cause much anguish to orchid growers. One species of orchid beetle is found in northern Australia. It mainly feeds on species and hybrids of *Dendrobium*, but it can also damage many other genera. If left unchecked, the larvae of this pest can ruin a year's growth and flowering in a few weeks. The natural occurrence of the Australian species is mainly in the tropical and subtropical parts of Queensland and northern areas of the Northern Territory. It can also become established in glasshouses further south.

This pest grows rapidly and many life cycles can occur within a couple of months. The eggs, which are fairly conspicuous and easily picked off by hand, are often laid on new growths, buds and flowers. The beetle larvae tunnel into new shoots usually causing them to die. They also feed actively on buds and flowers. Areas where the larvae are feeding become soggy and collapse. The larvae pupate in a mass of white, waxy material which resembles toothpaste. The adults, which are about 12 mm long, are shiny orange with four black spots on the wing covers. They graze the surfaces of leaves, flowers and seed pods. The leaves frequently develop white papery patches after this feeding. Orchid beetles are mainly active during the warm months of the year, although in the tropics they can be around for most of the year. They are a very persistent pest which can be very difficult to control. The usual recommendation is regular spraying with carbaryl while the beetles are active.

Caterpillars, Loopers and Grubs

A few caterpillars attack the leaves of orchids or more often young shoots and developing buds. Attacks by caterpillars are usually sporadic, but because of their voracious appetite, these pests can cause considerable damage in a relatively short time. The most persistent caterpillar which commonly becomes established in orchid collections is undoubtedly that of the Light Brown Apple Moth or Leaf Roller (*Epiphyas postvittana*). They are fleshy, green or pinkish, grow to about 1 cm long and form crude shelters by joining adjacent leaves together. When disturbed these caterpillars wriggle actively and frequently drop into the potting mix.

Other caterpillars which can cause sporadic damage to orchids include Painted Apple Moth (*Orgyia anartoides*), cutworms and loopers. Control in all cases is by squashing or spraying with materials such as bacterial spore suspensions or carbaryl.

Thrips

Thrips are tiny, slender insects which congregate in colonies and feed on plant sap leaking from areas which they damage. They are usually a minor pest of orchids and are mostly found feeding in the flowers. They are a very seasonal pest and in years with a warm dry winter they can build up into very large numbers and may then be a significant problem. Afflicted flowers develop brown patches or become distorted and turn completely brown. Frequently buds which have been damaged turn papery or appear dry. Thrips are more frequently a nuisance in terrestrials rather than epiphytes, although in bad years they will attack any flower. Control can be achieved by spraying with maldison or pyrethrum.

European Earwig (*Forficula auricularia*)

This common introduced pest is a significant nuisance in temperate regions and frequently becomes established in glasshouses and shadehouses. During the day the insects hide in dark crevices and emerge at night to feed. Favoured sites are in between the fibres of treefern slabs, under pads of sphagnum moss and in coarse potting mix. They feed actively on the root tips of epiphytic orchids, chewing out large lumps and destroying the very important growth apex. They can also damage new shoots and buds but this is less common. The numbers of this pest can be reduced by trapping in balls of crumpled newspaper which are collected each morning and burnt. Rubbish should be cleaned up and problem areas dusted with lime or naphthalene flakes. Baiting with snail baits and spraying with carbaryl are also effective measures.

Slaters or Woodlice

A common inhabitant of gardens which also finds the damp, humid conditions of glasshouses and shadehouses to its liking. This pest is very widespread in Australia but is especially prevalent in temperate regions. Slaters hide in crevices during the day and emerge at night to feed. They find excellent shelter between the fibres of treefern slabs, in the drainage holes of pots, in coarse potting mixes and in pads of sphagnum moss. While they mostly feed on decaying organic material (and in so doing assist in the breakdown of the potting mix), they can also damage orchids by feeding on the root tips. Continued feeding can cause stunting of the root system. Slaters are impossible to eradicate but commercial snail baits can provide some control. Clearing accumulated litter and spreading lime or napthalene flakes is a deterrent.

Millipedes

Millipedes inhabit damp situations and are often seen in glasshouses. They can be recognised by their narrow, jointed, black body with numerous small legs. When disturbed they coil characteristically and smell strongly when squashed. Millipedes feed by night and hide in crevices during the day. They take refuge in trash and cleanliness is of importance to keep numbers down. As with slaters, millipedes feed on the root tips of orchids and can cause stunting of a root system by continued feeding. Collecting with the aid of a torch at night and squashing is a useful means of reducing numbers. Commercial snail baits can also exert some control.

Cockroaches

The large black introduced cockroach, which is a familiar household pest, also feeds on succulent plant tissues and finds orchids to its liking. These obnoxious insects are especially prevalent in tropical regions (throughout the year), but are also common in the summer further south, flying on warm nights. They can cause damage to young shoots and are also fond of root tips. Control is by using traps or spraying with pyrethrum.

Black Crickets

Black crickets are usually seen during the late summer and autumn but they revel in warm conditions and can become established in glasshouses. They live in debris, crevices and under stones and pots. In some seasons they can be damaging to orchids, feeding on root tips and young growth. Control is by nightly inspections with a torch or by baiting with commercial snail preparations.

Locusts and Grasshoppers

These large pests cause sporadic damage to orchids during the summer months, particularly in tropical and subtropical regions. They chew large sections out of leaves, buds and flowers. Some types are shaped and coloured like a leaf and can be difficult to see. Control is by squashing or by spraying with carbaryl or maldison. These insects are often difficult to control because of their nomadic habits.

Rats and Mice

Although familiar as household pests, rats and mice can also be destructive of epiphytic orchids. They are commonly a nuisance in heated glasshouses during autumn and winter and are probably attracted into the structure initially by warmth. Mice find flowers and fleshy root tips tasty, whereas rats usually feed on fleshy pseudobulbs. Their feeding is obvious by the longitudinal grooves left by their teeth. Both of these pests can cause substantial damage in a short time and control measures should be initiated as soon as damage is noticed. Control of these pests is by trapping or by the use of commercial baits.

SUCKING PESTS

The following pests have mouthparts adapted for sucking. Their activities create a drying effect, and affected tissues become dull or papery, frequently associated with distortion if they occur on developing growths or flowers.

Aphids or Greenfly

New growths, developing inflorescences and buds are the targets of these common, persistent pests, which feed on epiphytes and terrestrials alike. They congregate in colonies and suck the sap from young tissue, resulting in wilting, distortion and bud drop. While feeding they excrete considerable quantities of sugary honeydew which attracts ants and is also an excellent substrate for the unsightly black fungus called Sooty Mould. Aphids are also a menace because they are vectors of plant virus diseases.

Many introduced species of aphids have become naturalised in Australia but only a couple are troublesome with orchids. The most persistent nuisance has yellow to orange immature nymphs and green adults. Most aphid young are born alive and are themselves capable of reproducing at an early stage. Thus they can build up in numbers very quickly unless control measures are taken. Hosing with a jet of water drowns many individuals and is an effective means of disrupting colonies. The feeding activities of predatory ladybirds and lacewings should be encouraged. Persistent infestation can be controlled by spraying with pyrethrum or maldison.

Spider Mites

A couple of species of these tiny eight-legged animals are among the most persistent and damaging pests of orchids and they cause growers considerable anguish. Unfortunately they find the warm conditions of glasshouses to their liking and once they become established are extremely difficult to eradicate. They spread easily from plant to plant and build up numbers at an amazing rate. Spider mites feed by sucking the sap and usually congregate in colonies on the underside of leaves. Attacks are worse on plants in dry corners, under benches or those that are neglected, badly in need of repotting or weakened from some other cause.

Fine webs are spun and these may cover the surface of the leaves, protecting the mites from moisture and sprays. Afflicted leaves appear dry and either turn a yellowish bronze colour with mottling or are silvery green and papery. Mite numbers are greatly influenced by climatic conditions and are most abundant in warm, dry weather. They often increase in numbers during the winter when humidity is low and growers water less frequently because their plants are not in active growth.

Two species of spider mite are significant pests of orchids but a hand lens or microscope is necessary for identification. Bryobia Mite (*Bryobia rubrioculus*) is a dull yellow colour and lacks any blotches. Two Spotted Mites (*Tetranychus urticae*) are also dull yellow but have two conspicuous blotches on the back. Winter populations are red all over and are then commonly known as Red Spider Mite.

Mite control is usually difficult and requires spraying with specific chemicals known as acaracides or miticides. Such chemicals include dimethoate (Rogor), tetradifon (Tedion) and difocol (Kelthane). Eradication of mites is well nigh impossible and because of their persistence, growers tend to be overzealous with sprays. This usually exacerbates the mite problem because excessive spraying quickly leads to selection of resistant strains. To reduce this build-up of genetic resistance, it is suggested that chemical sprays should be alternated. Mites generally resent free water and high humidity and jetting the underside of affected leaves is antagonistic to their activities. Predatory mites offer some hope of biological control and can be purchased from commercial firms. However, results with these in glasshouses have usually been less than satisfactory.

Scale Insects

There are many species of scale insects, all of which feed on plants, but only a few are active on orchids. Of these, however, some are among the most persistent and damaging of pests with which orchid growers have to contend. Scale insects conceal themselves beneath waxy shells and feed by sucking plant sap. Once attached to the sap stream by its mouthparts, the female scale insect loses its legs, becoming immobile and permanently attached to the plant. As the insect grows so its waxy shell is increased in size. The female insects spend all of their life beneath the shell, producing a mass of eggs before they die. Upon hatching, the young scales (known as crawlers) are highly mobile and disperse quickly over the plants before becoming attached. At this stage of dispersal they are most readily controlled.

Most species of scale are found in colonies. Ants are attracted to them and the unsightly fungus, Sooty Mould, commonly grows on their sugary exudates. Some details of scales which feed on orchids follow. One of the worst is called Woolly Scale, because severe infestations smother pseudobulbs and leaves with a dense mass of hairy scale coverings. It often starts in basal sheaths and spreads onto new growth and leaves, causing yellowing and premature death of growth buds. Soft Brown Scale (*Coccus hesperidum*) is sometimes found on epiphytic orchids and infestations can be persistent. The oval, flattish adults are mottled yellow-brown but the juveniles are often covered with white, waxy secretions. The adults of Nigra Scale (*Parasaissetia nigra*), have a leathery, oval covering which is black. This scale is often a persistent pest in orchid collections.

Localised infestations of scale insects can be squashed between the fingers or cleaned up with an old toothbrush dipped in methylated spirits, but persistent attacks may need to be sprayed. White oil at half strength combined with pyrethrum or maldison is usually effective. Note that white oil can damage orchid plants at full strength and it should never be applied when the temperature is above 25°C.

Mealy Bugs

Mealy bugs are soft, plump insects that are covered with a white, mealy powder which is water repellant. Commonly they congregate in small groups in sheltered, dry parts of plants such as on the underside of leaves, under sheathing bracts, in leaf folds, between crowded growths and on developing inflorescences. Their feeding activities often cause localised yellowing of tissue, and exuded waxy filaments are a common advertisement of their presence. Their feeding can weaken plants and result in the growth of Sooty Mould on their exudates.

Mealy bugs are a very persistent pest that is common in glasshouses. They not easy to control. Small infestations can be removed by dabbing with a cotton bud dipped in methylated spirits. Spraying with dimethoate (Rogor) or maldison (Malathion) can be effective.

Orchid White Fly

As suggested by the common name, the adults of these tiny insects are flies with white wings which are fine and gauzy. By contrast the legless young are immobile and are easily confused with scale insects. They are rounded, brown to black in colour and have a conspicuous marginal fringe of white, waxy filaments. White fly is a sporadic pest of glasshouse plants including orchids. They congregate in colonies and are often found on developing shoots, sheltering in or near the sheathing bracts. Their persistent attacks on weakened or debilitated plants can result in death. This is especially true of members of the subtribe Sarcanthinae. In plants of this group, they congregate in the leaf bases and along the margins where they overlap. Small infestations can be controlled by dabbing with a cotton wool bud dipped in methylated spirits. Spraying with pyrethrum may control persistent infestations.

DISEASES

Very few diseases affect orchids grown in Australia. Some are caused by virulent organisms but many are the result of stress or poor cultural techniques. Where good hygiene is practised and plants are observed regularly, the incidence of disease is greatly reduced. The presence of damaged tissue on a plant should not be neglected since such areas provide an ideal entry point for diseases. Cut surfaces should be dusted with lime or sulphur powder to ensure that the exposed area dries up and seals. Pest infestations can provide entry points for disease organisms. Fungal spores are omnipresent but localised infections can be made worse by the presence of decaying litter. All prunings, weeds and other litter should be removed from the glasshouse and not be left and allowed to decay. Air movement is of major importance since fungi build up rapidly under stagnant conditions. If a disease is noticed then control steps should be taken immediately because they can spread very rapidly.

Plant Viruses

Plant viruses are minute particles of nuclear tissue which live and reproduce within plant cells. Their presence disrupts normal growth and can cause significant distortion and malformation. The most severe effects are shown when they build up in very large numbers (such as in old plants); when they are present in stressed plants (such as those which are nutritionally deficient); when different virus species occur together in the plant cells. The latter situation can cause severe disruption of growth and the appearance of symptoms which may be different from those which either virus produces separately. Typical symptoms caused by viruses include irregular mosaic patterns in tissues, necrotic streaks and spots, reduction in size and distortion of leaves and colour breaks in flowers. Viruses are spread by unclean cutting tools and sucking insects such as aphids and leafhoppers. The effects of virus are perpetuated by vegetative propagation. Virus infection cannot be controlled by spraying and any suspected plants should be isolated from the main collection or destroyed by burning.

Several viruses are known to occur in cultivated orchids. The most prevalent include Cymbidium Mosaic Virus, Cymbidium Necrotic Ringspot Virus, Odontoglossum Ringspot Virus and Tobacco Mosaic Virus.

Root Rots

Orchids, particularly epiphytes, can suffer from the depredations of a number of vigorous soil-borne fungi. These fungi, which include species of *Pythium*, *Phytophthora*, *Fusarium*, *Pellicularia* and *Rhizoctonia*, attack the root system causing the plants to wilt, followed by leaf yellowing and eventual death. Root rot can be disastrous in seedlings especially if they are being grown in community pots. Potting mixes can become infested when mixing or pots may be contaminated by rain splash or the fungi can enter through the drainage holes if the pots are in contact with soil. These fungi are more devastating if the potting mix is incorrect (especially if drainage is inadequate) or if the plants are overwatered or weakened by being grown in too much shade. Afflicted plants of terrestrial species develop a watery brown rot which spreads quickly and causes death. Control is to identify and correct the problem; that is, improving factors such as drainage and air movement, repotting if necessary, protecting plants from heavy rain and drips and growing them on benches above the soil. Drenching affected pots with furalaxyl (Fongarid) may inhibit the spread of the fungus in the short term.

Bacterial Soft Rot

A very damaging pest that causes much anguish to orchid growers, especially in the tropics. Known as *Pseudomonas cattleya*, it is capable of very rapid spread and can destroy a plant within a couple of days. First symptoms are usually a circular, water-soaked area which quickly increases in size. Infected tissues become soggy and release exudates which are full of bacteria and can spread to adjacent plants. Infected plants must be isolated and treated immediately the disease is noticed. Affected organs should be cut off well in front of the zone of infected tissue and the whole plant is soaked in Natriphene. Regular examinations will be necessary to ensure that the infection has been curtailed.

Grey Mould (*Botrytis cinerea*)

A common fungus of glasshouses, Grey Mould attacks damaged leaf tissue and flowers in still, humid conditions. It forms a mass of fungal mycelium which becomes very prominent when the spherical grey spores are produced. These are particularly noticeable when wet. Grey mould can attack both terrestrials and epiphytes. In terrestrials it usually enters tissue damaged in some way such as by pests, heavy watering, rain splash or drips. In epiphytes it usually starts on spent flowers, fallen leaves and such and can spread to healthy tissues. Control is mainly by hygiene, reducing plant damage and improving aeration. Spraying with benomyl (Benlate) or Zineb is usually effective at controlling sporadic outbreaks.

Leaf Spots and Rots

Various fungi attack the leaves of orchids causing spots, blotches and sometimes rots. Often the spots start on the undersides of the leaves and are not noticed until the upper surface becomes affected. Frequently the spots are surrounded by a halo of pale green or yellow tissue. Various leaf rots can be a major problem of terrestrial orchids especially in tropical and subtropical regions. They are worse on plants grown in the open and attacks often follow the damage caused by heavy rain or sucking pests.

Anthracnose Spot is common in tropical and subtropical regions. The leaves develop brown bands and concentric rings and flowers are dotted with brown to black raised lumps or pustules. Control is by spraying with Mancozeb or Zineb.

Fusarium Spot or Fusarium Wilt is a serious disease which can start as a leaf spot and spread into the vascular system, blocking the vessels and causing wilting, yellowing and death. Frequently white fungal threads and pinkish spore masses can be seen as the infection develops. This fungus can also cause bud drop on developing inflorescences. Infected plants usually die from this fungus unless drastic action is taken at an early stage. Spraying with Mancozeb and Zinc will control bud drop.

The leaf tips of *Cymbidium* can be damaged by a fungus known as brown rot. This is generally a minor problem that can be controlled with sprays of copper oxychloride.

6

Housing

A small collection of orchids can often be successfully grown under the shelter of trees, on protected benches, verandahs or even in modified areas indoors. Orchid collections, however, have a habit of expanding and for optimum growth, substantial collections must be housed in some way. The alternatives are a solid structure such as a glasshouse or greenhouse or an open, airy shadehouse, lathehouse or bushhouse. The choice depends on factors such as the type of orchids being grown (in particular how much the climate must be modified for their successful growth), the prevailing climate and the exposure of the site. In tropical areas a wide range of orchids can be successfully grown with scant protection and most growers choose a simple shadehouse type structure. In temperate regions a shadehouse will be suitable only for cold-hardy species and the protection of a glasshouse will enable a much greater range of species to be grown successfully. If that glasshouse is heated then the range will be further increased. Whatever the choice, a suitable structure must protect the plants from climatic extremes and encourage their growth by maintaining warmth and humidity.

GLASSHOUSES AND GREENHOUSES

Conditions within a glasshouse are generally warmer than outside and the humidity higher, thus improving orchid growth and allowing a greater range of genera and species to be grown. Orchids which are grown in a glasshouse also have a longer growth cycle because of an earlier start to the season and a later finish. Frosts cause severe damage to all but the hardiest of orchids and glasshouses provide good protection from such climatic extremes. Glasshouses are commonly encountered in temperate and colder subtropical regions, but they also have a role in highland tropical regions where winters are usually cold.

Size

Whatever the size of the glasshouse built, it usually ends up too small. Size depends largely on finances but it should be pointed out that small glasshouses heat up and cool down much more quickly than larger constructions and these fluctuations can be detrimental to the orchids. The plants are also crowded and because of this restriction, pests and diseases may become a real problem.

Siting

Glasshouses should be situated where they receive maximum winter sun. Shading by trees can be beneficial in summer by reducing the upper extremes of temperature, but shading in winter can be detrimental or even disastrous. As a general rule, orchids require good, bright light during winter even though most species may be dormant at this time. Winter shading also reduces the limited warming effects of the weak winter sunshine and will increase fuel bills dramatically.

Orientation

Despite much speculation about glasshouse orientation, studies indicate that orientation is only of significance at high latitudes. Thus in southern Tasmania it is important to align the glasshouse so that it presents its translucent surface at right angles to the sunlight. Over most of the mainland however, the total amount of light received in a glasshouse is largely independent of its orientation.

Construction

Glasshouses can be purchased from commercial firms in kit form, commonly with aluminium or steel frames. These kits are usually of a basic rectangular design with a gable roof and are relatively simple to erect. Some designs are more suited to temperate areas rather than the tropics and factors such as provision for ventilators are quite variable between manufacturers, so it is a good policy to look at the suitability of a completed article before buying a kit. Glasshouses can be constructed by a handyman from scratch. Home-built units can be designed to fit into a particular situation and are usually based on a timber frame. Pressure-treated softwood is ideal for the frame since it is long-lasting, light and easy to work. Glasshouses require a strong frame because the cladding materials (especially glass) are generally heavy. The frame should not be excessively bulky and should not cause more than 8% shading. Glasshouses must usually conform with local building codes.

Shape

Glasshouse shapes are variable and are often influenced as much by aesthetic factors as by practical considerations. The construction should be kept as simple as possible and shapes and profiles that have proven themselves over years should be chosen rather than complex types. A major consideration is to obtain optimum efficiency from the bench area since this is where most of the orchids will be grown.
Paths can waste up to 40% of the floor space in poor designs. A simple technique to overcome this is to draw a floor plan of the glasshouse to scale on graph paper and juggle a series of bench cut outs to scale until the most efficient use of space is found. Pathways should be at least 70 cm wide to allow for easy access.

Cladding

A range of materials is available to cover a greenhouse frame, and the choice will largely depend on cost and availability. Glass is the traditional covering material for a glasshouse but its position is being threatened by modern plastic materials. Glass has the disadvantages of being heavy and therefore requiring the support of a strong frame and fragile. However it has the best light transmission properties and does not deteriorate with age. Rigid plastics and fibreglass are excellent modern alternatives, being much lighter, easier to handle and less fragile. However, they usually have a shorter life, especially if cheap grades are used. While they are often available in a range of colours, only opaque white or clear are suitable for orchid growth.

Ventilation

Adequate ventilation is of major importance in successful orchid culture. Orchids require a free flow of fresh air and resent stale conditions. Adequate ventilation also helps to cool the glasshouse during hot weather and assists with disease reduction. Ventilators should be situated near the base of walls and in the roof so that an upward flow of air is achieved. Small fans and polythene ducts can also help to distribute air evenly around the glasshouse and are very useful for air circulation when the ventilators are closed.

Shading

A glasshouse must have supplementary shading over summer, otherwise the orchid plants can become excessively bleached or burnt. The simplest shading technique is to paint the glass with a white commercial glasshouse paint or limewater with some linseed oil added. This is a messy and dirty job and shadecloth materials supported on a frame 15-30 cm above the glass provide an excellent modern alternative. The shadecloth should not be laid flat on the glass as this causes extra heat to build up in the glasshouse.

Humidification

A congenial atmosphere builds up in a successful glasshouse, and an experienced grower can immediately tell from its feel, whether the plants need watering or the humidity needs enhancing by hosing the floor, walls, etc. This atmosphere is created mainly by humidity with a contribution from the plants. It can be enhanced by strategies such as lining the walls with polythene sheeting and the use of mist sprays. In glasshouses with concrete floors the humidity tends to fluctuate. Watering the concrete each morning is a simple means of enhancing the atmosphere.

Heating

Heating a glasshouse increases the range of orchids which can be grown, but at a cost. Heating equipment can be expensive and there are annual running costs to be considered. Some aspects of glasshouse construction can improve heating efficiency. Cladding materials greatly influence heat loss. Thus brick walls are more efficient at conserving heat than single sheets of materials such as asbestos. A double glazed roof loses about half the heat of a single sheet construction but is more difficult to construct and usually cuts down light transmission. Lining the inside of a glasshouse with polythene sheeting reduces the heat requirements by 35%.

The best heating system for orchids is the traditional method of passing hot water through pipes. This produces gentle, humid heat but requires the use of a boiler and can be costly to set up. Systems which heat the air directly are cheaper but result in drier air which must be humidified in some way. Mist systems provide a simple answer to this problem. The heat is distributed evenly through the glasshouse by means of ducts or polythene sleeves. The use of heat banks or heat sinks can make a significant difference to the winter temperatures of an unheated glasshouse. Simple systems such as rock piles or stacking large containers of water under benches can effectively avoid damage from light frosts.

SHADEHOUSES, BUSHHOUSES AND LATHEHOUSES

In general these structures provide shade and maintain higher humidity than outside while still giving excellent aeration. The humidity also fluctuates much less than outside. Shadehouses can have a solid roof (which reduces damage caused by rainfall and drips), or be clad with modern shadecloths. Such structures provide minimal frost protection, especially during heavy frosts. They are commonly used in tropical and subtropical climates to provide protection from hot sun but can also be successfully employed in temperate regions for the cultivation of hardy orchids.

Siting

Shadehouses should be situated where they receive maximum winter sun. This is especially important in temperate regions because if placed in a shady location the effect of winter cold is aggravated and the range of species which can be grown is correspondingly reduced.

Construction

Shadehouse frames are commonly constructed from pressure-treated softwood or galvanised pipe. The frame is relatively simple to construct and is readily covered with some shading system. The modern grower is fortunate in having a range of woven or knitted fabrics from which to choose. For orchids, a cloth which provides 50% shade is satisfactory in most areas. These shadecloths are generally light and durable but aesthetically do not appeal to everyone. Some growers prefer regularly spaced wooden slats or lathes which provide dappled shade. If these are painted in brownish or greenish tones they blend in well with the surrounds and can be an attractive addition to the outside decor.

The roof of a shadehouse can be made from solid materials such as fibreglass or plastic. Such a solid cover will give additional protection from severe cold and eliminate the damage caused by heavy rain and drips from the shadecloth. Wind can cause damage to plants in a shadehouse and if one side of the structure is exposed to strong winds then it may be advantageous to build a solid wall on this side.

7

Orchid Propagation

Simple techniques of propagation can be employed by orchid growers to increase their collection. These methods can be carried out with a minimum of equipment and when successful give the grower much satisfaction. More detailed techniques of propagation such as seed sowing and meristem culture are also possible but they require elaborate equipment and considerable time. Nevertheless dedicated growers have mastered these techniques and use them as a matter of routine for propagation of species and to raise new hybrids. Vegetative techniques of propagation produce progeny identical to the parents whereas seedlings are often variable.

DIVISION

Large clumps of sympodial orchids can be propagated readily by division. The optimum time for dividing a plant is when the new growths are just developing from the base of the mature pseudobulbs, although some growers prefer to cut the rhizomes before this and divide the clumps when the new shoots appear on the divisions. Divisions which have new shoots establish quickly as separate plants because strong new roots are produced from the base of each growth as it matures.

As a general rule applying to a range of species, orchid plants should not be divided excessively. Divisions comprising two or three pseudobulbs together with a new shoot are satisfactory for many species. Very vigorous orchids can be cut at each bulb for propagation purposes but such plants will not flower for a number of years. Species of *Odontoglossum* are notorious for not shooting from mature divisions and for these orchids the rhizome is severed immediately behind the most recently matured pseudobulb. Both sections should produce new growths and are potted separately when sufficiently mature. Some orchids such as *Bifrenaria harrisoniae* and *Coelogyne pandurata* are sensitive to disturbance and suffer a setback when divided. Such species are best left until division becomes absolutely necessary and then substantial clumps are retained.

Dividing Technique

1) Either precut the rhizomes into sections supporting two or three pseudobulbs or cut the rhizomes at the time of repotting. First knock the plant out of its container and shake free

of all potting mix. Pre-sterilise knife or secateurs (sharp in both cases) by passing the blade through a flame or dipping in alcohol. Cut surfaces should be sealed with lime or powdered sulphur.

2) Cut all dead roots from the division.

3) Pot into a container using suitable potting mix. Allow ample room when potting for forward growth from the division. Pot the plant to its original level, firm the mix and water well.

4) Pieces which are very small or have no roots may require specialised treatment. Potting into a small container of sphagnum moss is usually successful, especially if a bottom-heat unit is available. Once they have made a new growth and new roots, these divisions can be potted normally.

BACKBULBS

Backbulb is a term used for an old, but still live, pseudobulb which has shed its foliage but still retains dormant buds. For some orchids, such as species of *Cymbidium*, these backbulbs can be separated individually and will produce a new shoot from near the base. In others such as *Cattleya* and *Laelia* they should be retained in groups of three to five and there is a good chance a new shoot will be produced from one backbulb in the group.

After separation from the plant by cutting the rhizome, all roots should be trimmed from the backbulb(s) and old sheathing bracts removed to reveal pests such as scale or mealy bug. These can be squashed and the pseudobulbs are then buried for half their length in a small pot of sphagnum moss and placed on a bottom heat unit. When the new growth is substantially developed with good roots, the plants can be potted separately.

AERIAL GROWTHS OR KEIKIS

Keiki is the Hawaiian word for baby and it has been adopted by orchid growers as an affectionate term for aerial growths which arise spontaneously from the apical nodes of many orchids. These growths develop as a vegetative shoot which matures into a pseudobulb with leaves and roots. Aerial growths can produce new shoots from the base and have the ability to form a separate plant if removed from the parent growth.

Keikis are an excellent means of propagation but they should not be removed from the parent plant until mature and with good root growth. Overlooked aerial growths develop into miniature clumps while still attached to the parent plant. These establish very quickly when removed and potted. When mature, aerial growths can be broken easily from the pseudobulb and do not need to be cut. They should be potted into a small pot until well established. Sphagnum moss is an excellent potting medium, especially if the root system is poor. Orchids which produce aerial growths include *Cyrtopodium punctatum*, *Dendrobium anosmum*, *D. bigibbum*, *D. discolor*, *D. heterocarpum*, *D. nobile*, *D. pierardii*, *D. X superbiens*, *D. superbum*, *Epidendrum* spp., *Malaxis* spp., *Oncidium* spp. and *Thunia marshalliana*.

Species of *Pleione* reproduce prolifically by plantlets produced from the top of their declining pseudobulbs. These fall off naturally and establish in the pot or else they can be removed and potted when sufficiently mature.

Some vandoid orchids produce aerial growths from the lower nodes of their flower stems. These develop into sturdy plantlets and can be easily separated when sufficiently well developed. Growers can enhance plantlet production of these orchids by applying a small quantity of growth hormone (indole acetic acid (IAA) or naphthalene acetic acid (NAA) in lanoline paste) to the lower nodes after removal of the flowers. The commercial product, known as keiki paste, produces similar results. Certain species of *Phalaenopsis* (particularly *P. lueddemanniana*), *Aerangis kirkii*, *A. rhodosticta* and *Vanda* (especially *V. cristata*) can be propagated by this means.

MONOPODIAL CUTTINGS

Monopodial orchids which have a tall growth habit can be propagated by cutting the plant into sections. Only vigorous, strongly growing plants should be treated this way and the cuttings should not be too small. The best time for this type of propagation is in summer when the plants are in active growth. The stems should be cut below a strong root leaving at least three pairs of healthy leaves on the basal part. All cuts are sealed with lime or powdered sulphur. The top cut can be potted as a separate plant and should establish strongly. The bottom, already established part of the plant should produce one or more strong new shoots fairly quickly.

Monopodial orchids which produce strong growths from the base are easily propagated. When sufficiently well-developed with their own roots, these growths can be severed and potted separately.

SOFTCANE PROPAGATION

Soft cane species and hybrids of *Dendrobium* which produce flowers along the length of their canes can be propagated by a simple technique. A mature cane which has dropped its leaves, but not yet flowered, is severed at the base and cut into sections each containing two nodes (the swollen humps represent a node). The cuts are sealed by dipping into lime or sulphur and the pieces are laid in sphagnum moss in a warm glasshouse or propagation unit. Shoots develop from some of the nodes and after six to twelve months they are sturdy little plants with their own roots. The best time to carry out this simple technique is in spring, before the buds at the nodes have begun to swell. Suitable species include *Dendrobium anosmum*, *D. heterocarpum*, *D. nobile*, *D. pierardii*, *D. superbum*, *D. transparens* and *D. wardianum*.

FLOWER STEM PROPAGATION

Several species of *Phaius* produce fleshy flower stems that remain green and viable on the plant even after the seed has been shed. *Zygopetalum mackaii* sometimes has a similar habit. Although they eventually wither and die, these stems can be used for propagation purposes if cut from the plant while still fresh and green. After cutting into lengths 10-15 cm long, these are laid in sphagnum moss in a warm glasshouse or propagation unit. Shoots develop from some of the nodes and after six to twelve months these have grown into plantlets with their own roots. When well developed these plantlets can be potted separately.

SEED PROPAGATION

The methods required to raise orchid seeds under sterile conditions are time-consuming and complicated when compared with raising vegetables or nursery crops. The procedure however, is easily learned and any problems can be ironed out after a few practice runs.

Cleanliness is essential throughout the whole sowing operation and this factor should be considered at each stage. Contamination can occur at any point and will result in the germination and growth of fungi which will destroy the orchid seedlings. Bacteria can also be contaminants. Contamination will usually show up within a few days of sowing and infected flasks must be discarded.

Preparation and Sowing Site

A special, easily sterilised sowing area greatly reduces the risk of contamination and makes the whole sowing procedure much easier. Benches with a smooth surface or a simple cabinet fitted with armholes are more than useful. A lamina flow cabinet provides the ultimate sterile work area. Surfaces of the work area should be sterilised with alcohol (methylated spirits is satisfactory) or bleach solution prior to use. All equipment including sowing flasks, stoppers, forceps, platinum loops, etc, should be sterilised before being placed in this sanitised zone.

Sowing Media

Orchid seeds are sown on an agar medium to which has been added nutrient salts and sugar. This is a basic formulation and it can be varied by changing the type of nutrient salts or by adding compounds such as vitamins, growth regulators and fruit products (ripe banana and green coconut milk). There are almost as many different media as there are orchids and growers often adopt a particular formulation as their favourite.

The original formula for orchid seed germination, developed by Lewis Knudson in the 1930s, is still used by many growers and its ingredients are presented in Table 1 below. Also included are two other formulations, Vacin & Went which is often used by research workers and another with which the author has had good success (see Table 3). The latter is based on Knudson's media but with some useful additives.

Table 1 – Knudson's Orchid Formula 'C'

Chemical	Symbol	Quantity
Sucrose	$C_{12}H_{22}O_{11}$	20 g
Agar		15 g
Calcium nitrate	$Ca(NO_3)_2.4H_2O$	1 g
Ammonium sulphate	$(NH_4)_2SO_4$	500 mg
Monobasic potassium phosphate	KH_2PO_4	250 mg
Magnesium sulphate	$MgSO_4.7H_2O$	250 mg
Ferrous sulphate	$FeSO_4.7H_2O$	25 mg
Manganese sulphate	$MnSO_4.4H_2O$	7.5 mg
Water	make up to 1 Litre	

Table 2 – Vacin & Went Medium

Chemical	Symbol	Quantity
Sucrose	$C_{12}H_{22}O_{11}$	20 g
Agar		9 g
Potassium nitrate	KNO_3	525 mg
Ammonium sulphate	$(NH_4)_2.4H_2O$	500 mg
Monobasic potassium phosphate	KH_2PO_4	250 mg
Magnesium sulphate	$MgSO_4.7H_2O$	250 mg
Iron Chelate		42 mg
Manganese sulphate	$MnSO4.4H2O$	7.5 mg
Tricalcium phosphate	$Ca3(PO4)$	1 mg
Water	make up to 1 Litre	

Table 3 – Fetherston's Orchid Formula

Chemical	Symbol	Quantity
Sucrose	$C_{12}H_{22}O_{11}$	20 g
Agar		12 g
Calcium nitrate	$Ca(NO_3)_2.4H_2O$	1 g
Ammonium sulphate	$(NH_4)_2 SO_4$	500 mg
Monobasic potassium phosphate	KH_2PO_4	250 mg
Magnesium sulphate	$MgSO_4.7H_2O$	250 mg
Iron chelate		42 mg
Manganese sulphate	$MnSO_4.4H_2O$	7.5 mg
Vitamin C		125 mg
Vitamin B_1		2 mg
Nicotinic acid		1 mg
Indole acetic acid		1 mg
Coconut milk		200 mL
One ripe banana (blended)		
Water	make up to 1 Litre	

Preparation of Sowing Media

The required chemicals can be purchased from chemists or chemical supply companies. The concentrations of each chemical are critical and they must be weighed accurately on a sensitive balance. This can often be arranged through a chemist.

Preparation of the medium is best approached step by step. For any medium warm about 800 mL of water and add each of the ingredients, stirring gently until it is dissolved before adding the next. The agar is added last and must be thoroughly stirred as it can be slow to dissolve. It may be necessary to heat, but not boil, the medium to ensure solution of the agar. Make up to one litre with warm water.

A small sample of the mix is removed and tested with a pH indicator (colour solution or indicator paper) for its acidity or alkalinity. The pH of the mix should be 5-5.2 and, if not, it is adjusted by adding acid or alkali solution. If too acid (below 5), a drop of 0.1 Normal potassium hydroxide is added to the bulk solution, stirred and retested. If too alkaline (above 5.2), adjustment is made with drops of 0.1 Normal hydrochloric acid. Usually the pH of the bulk solution changes rapidly with the addition of small quantities of acid or alkali, so only add one drop at a time.

When cool, the sowing medium sets to the consistency of jelly. It retains a thin film of water on the surface in which the orchid seeds germinate.

Seed Flasks

Orchid seeds are usually sown in glass flasks of the Erlenmeyer type, but growers may successfully use cream bottles, whisky bottles or screw-capped jars. Whatever the shape used, it must be of glass and be capable of being sterilised.

Screw-capped jars are sealed with a screw-on lid but other types of flasks must be sealed with a tight-fitting rubber stopper. A hole is bored through the centre of the stopper and a piece of glass tubing is inserted. This is bent near the middle (or is fitted with a piece of plastic tubing) and is plugged by stuffing with a wad of cotton wool. Thus filtered air can reach the orchid seedlings inside. The cotton wool may be treated with a solution of mercuric chloride or copper napthenate to kill any contaminating spores which blow in. A similar, filtered ventilation system must be inserted in the centre of a screw-on lid and sealed with a cement such as Araldite.

Flasks, bottles and stoppers must be thoroughly cleaned before use. The prepared agar is dispensed into the flasks, using a funnel to avoid splashing. The stopper is inserted and the flasks are then ready for sterilisation.

Sterilising the Prepared Flasks

The orchid medium provides an ideal substrate for the growth of fungi and bacteria. These grow rapidly and smother the orchid seeds. The only successful method of control is complete sterilisation of the flasks and media prior to sowing.

The stoppered flasks containing the agar medium are best sterilised in an autoclave; however, a pressure cooker can also be used successfully. Exposure for twenty minutes at 15 psi is usually sufficient to ensure sterilisation. After removal from the autoclave, the agar can be allowed to set in a horizontal position or on a slant (which increases the surface area). When cool, the flasks are ready for sowing.

Seed Sterilisation

Orchid seed is often contaminated with fungal spores and must be sterilised before sowing. Fresh solutions of calcium hypochlorite (10 g calcium hypochlorite with 140 mL water) or 0.5% sodium hypochlorite (everyday household bleach) are satisfactory. A small quantity of seed is soaked in the sterilising solution for periods of between five and twenty minutes. Shaking a mixture of the seed and bleach solution in a small bottle is a simple method of sterilisation. After sterilisation, the hypochlorite solution is drained off and the seeds are washed with a few changes of sterile water. The seeds of some orchid species may be hygroscopic and are very difficult to wet. The addition of a drop of household detergent can make such seeds easier to wet.

Resistant fungal spores may be present in some batches of seed and these will germinate later, contaminating the flask. If this is known to be a problem in a batch of seed then pre-soaking for a day in a weak sugar solution (two teaspoons to a cup of water) is the answer. This sugar mix will induce the fungal spores to germinate, after which they are readily killed by the bleach solution.

Seed Sowing

After the final wash, the excess sterile water is drained off leaving a concentrated solution of orchid seed. This can be sown directly into the flasks with a sterile eye dropper (four drops per flask) or pipette, or transferred with a sterile wire loop. Always remember that orchid seed is tiny and it is easy to sow too heavily, resulting in weak, crowded seedlings. It is far better to sow a small amount of seed rather than an excess.

Green Pod Sowing

Orchid pods can be removed from the flower stem when about two-thirds developed and the immature seeds can be germinated. To obtain the seeds the pod is first sterilised by dipping in bleach solution and then cut open. The seeds can then be induced to germinate if sown onto nutrient agar in sterile conditions. The technique of using green pods simplifies orchid propagation by seed since only the outside of the pods needs to be sterilised. It is used mainly for epiphytic species.

Germination and Development

After sowing, the flasks are placed in a warm, well-lit position and observed regularly. If the seeds are viable, swelling occurs rapidly and within six weeks small green, globular bodies begin to form. These are protocorms and the first leaves are formed three to six months after sowing. The young plants continue to develop steadily and roots appear within ten to twelve months. When the plants are well-developed they can be removed from the flask and grown in community pots.

Deflasking Seedlings

When ready for transplanting, the orchid seedlings are washed from the flask using warm water. This dissolves the agar and all traces of agar solution should be removed from the plants by repeated washing. They should then be completely immersed for a few minutes in a mild fungicide solution prior to potting.

After washing in fungicide, the young plants are placed close together in the bottom of a container. Suitable containers include large pots, ice cream containers or 9 L buckets. The bottom of the container is covered with a 5 cm thick layer of live sphagnum moss, preferably chopped into short lengths. The plants are arranged about 3 cm apart and the roots are covered lightly with moss. The whole pot is then dampened and covered with a sheet of glass to maintain humidity. The container is placed in a warm, shady situation and the plants are inspected regularly for fungus attack. Green root tips are a sign of healthy growth. After two to four weeks the glass can be removed and the seedlings allowed to develop further. When sufficiently developed they can be potted as separate plants.

TISSUE CULTURE

Tissue culture (or meristem culture) is a modern technique of micro-propagation whereby exeedingly small pieces of plant tissue (termed explants) are induced to multiply and then eventually differentiate into small plants. The technique is excellent for producing large quantities of plants in a short time. The progeny will be identical to the parent from which the explant was taken, and hence the technique is mainly of interest to commercial growers who produce hybrids or cut flowers. The progeny will also be free of diseases, including viruses. Many orchids have been propagated successfully by tissue culture techniques, although difficulties have been encountered with some genera. Spectacular results have been obtained with *Cymbidium* species and cultivars.

Equipment

Tissue culture propagation must be carried out in a laboratory using sterile techniques and culture media containing nutrients and hormones. The growing flasks are held in cabinets where light (photoperiod) and temperature can be controlled.

Technique

The starting tissue is usually a shoot meristem obtained from a dormant bud. The pseudobulb containing the bud must first be removed from the plant and sterilised completely. The bud is then cut out using a sterile scalpel and trimmed down to the growing point or meristem. This is introduced into a sterile flask of nutrient culture which is then kept agitated. If successful the explant develops and the tissue proliferates into a mass of undifferentiated callus cells. This mass of cells can be cut into pieces which will themselves proliferate when transferred to new culture flasks. By continuing this multiplication process the amount of tissue in culture can be built up rapidly. Multiplication factors of fifty to one hundred times are not uncommon.

When sufficient tissue has been propagated, the addition of plant hormones to the media will cause the callus to differentiate into small plants, each with roots and leaves. Naphthalene acetic acid (NAA), or a close derivative, is the hormone commonly used to induce differentiation. When the new plants are sufficiently developed they can be deflasked using the techniques outlined in seed propagation.

The Orchids

This section details some 230 species of orchids, each accompanied by a colour plate. Commonly cultivated and flamboyant species are well represented but also included are many lesser-known species and those mainly of botanical interest. Up-to-date names are used wherever these are known and for recent and important changes a cross-referenced entry is included in the index. An entry for each genus provides details of its size and distribution and summarises suitable general cultivation techniques. As well as touching on points of general interest, the index dealing with each species highlights features of its natural growing environment which will influence its cultivation. At the bottom of each species entry is a simple table which is designed to aid cultivation.

The following parameters have been used as a guide.

TEMPERATURE

Cool	Minimum temperature 5°C
Intermediate	Minimum temperature 10°C
Hot	Minimum temperature 16°C
Very Hot	Minimum temperature 22°C

SHADE

10-30%	Bright light lovers
30-50%	Medium shade
50-70%	Shade lovers

IN/ON

Preferred growing container or substrate.

HUMIDITY

30-50%	Low humidity
50-70%	Medium humidity
70-90%	High humidity

AIR FLOW

A scale of 1 to 5 with 5 being a strong, unimpeded air flow.

ACANTHEPHIPPIUM Blume

(from the Greek *acantha*, thorn; *ephippion*, saddle; the callus of the labellum has a fanciful resemblance to a saddle)

A genus of about 15 species of orchids distributed from tropical Asia to Fiji. Most species grow as terrestrials but epiphytes are also known. The plants are fairly large with broad leaves and elongated pseudobulbs which are often grooved. Inflorescences arise with the new growth and the large, flask-shaped or jug-shaped flowers have fleshy segments. In this genus the column has a long basal foot together with an extension to which the labellum is attached. Those species so far tried have proved to be easy to grow but in general orchids of this genus are only found in the collections of enthusiasts.

Cultivation: These orchids have strong, coarse roots and prefer a sizeable container in which a good root system can develop. Mixes containing some good loam, coarse sand, chopped fibre and fine particles of bark are suitable. Plants are evergreen but generally cease growth over winter and should then be kept on the dry side. During the growing cycle they require warmth, moisture and respond to fertilisers. Some growers top the pots of these orchids with old cow manure.

Acanthephippium striatum

Acanthephippium striatum Lindley

(finely striped)

Native to Nepal and the Himalayas, this species grows as a terrestrial on sheltered slopes at about 1000 m elevation. The thick fleshy flowers vary somewhat in colour from white to pale yellow, but all variants are characteristically marked with red stripes and streaks. The flowers have a heavy, spicy perfume which is most noticeable on warm days. An easy species to grow and flower, this is a good orchid for the novice. Leaves will burn or bleach if exposed to sun or excessive light. For this reason some growers move the plants to a shadier position on the bench or even underneath.

Temp	Shade	In/on	Humidity	Air Flow
Int	50-70%	Pot	50-70%	3-4

ADA Lindley

(after Ada, historical Greek woman)

This very small genus consists of two species, both of which are epiphytes endemic to Colombia. Although fairly closely related to *Odontoglossum*, species of this genus have brightly coloured flowers which are tubular at the base and with narrow spreading segments. Both species have bright orange flowers clustered on wiry, often arching racemes.

Cultivation: Both species of *Ada* are relatively easy to grow and are good orchids for beginners. They require year-round moisture, high humidity and shade. They are best grown in small pots in fresh sphagnum moss or a fibrous mix. Growth continues throughout the year but is slower during winter.

Ada aurantiaca Lindley

(orange-yellow)

Of very easy culture, this highly decorative orchid is an excellent one for beginners. Originating from more than 3000 m altitude in the Andes of Columbia, this species is quite cold tolerant and does not require heat to overwinter in temperate Australia. Plants grow well in similar conditions to those needed for species of *Odontoglossum*. While they can be grown on slabs, their upright growth habit makes them very suitable for pots. With their fine roots they need a potting mix which has small to medium-sized particles.

Temp	Shade	In/on	Humidity	Air Flow
Cool	30-50%	Pot	50-70%	5

AERIDES Lour.

(from the Greek, *aer*, air; *eides*, resembling; in reference to the epiphytic habit)

A genus of elegant orchids consisting of about sixty species which are distributed in India, Burma, South-east Asia, Japan, Malaysia and Indonesia. All grow on trees or rocks in humid situations and in bright light. *Aerides* have a similar growth habit to species of *Vanda*, but plants can be readily distinguished by brown suffusions on the old stems. The drooping racemes of colourful crowded flowers have earned them the

Ada aurantiaca

name of Fox Tail Orchids. In most species the waxy, long-lasting, heavily fragrant flowers are white but often they have colourful blotches and suffusions. The stems of most young plants of *Aerides* are erect but as they grow and elongate the stems becomes horizontal or even pendent with the apex curving upwards. In nature large-growing species have been recorded with stems more than 6 m long. Plants do not achieve these dimensions in cultivation. Basal growths and branches are a useful means of propagating *Aerides*. Most species of *Aerides* grow extremely well in the tropics and are often used to decorate gardens.

Cultivation: *Aerides* are popular with orchid growers and are often grown in hanging pots or slat baskets. In South-east Asian countries they are commonly tied to teak baskets without any potting mix. This ensures the excellent drainage and aeration which these orchids require around their roots. In Australia growers use coarse materials such as large pieces of bark, charcoal, scoria and terracotta shards. All species need plenty of warmth, high humidity, bright light and abundant air movement. Most growth occurs during the summer months but plants continue slow but steady growth over winter. Many growers hang plants close to the glass at this time to achieve as much warmth and light as possible. These orchids are gross feeders and lumps of old animal manure and liquid fertilisers are highly beneficial.

Aerides fieldingii

Aerides falcata Lindley
(sickle-shaped, in reference to the leaves)

A large coarse orchid which may grow to about 2 m tall and with curved, leathery leaves each about 30 cm long. Plants are transformed, however, by the arching or pendulous racemes which bear handsome, colourful flowers crowded along their length. Well grown plants flower freely and provide an impressive floral display. An excellent garden plant for the tropics. In nature, the species grows on trees and rocks in India, Burma, Thailand and Laos.

Temp	Shade	In/on	Humidity	Air Flow
Hot	10-30%	Pot	30-50%	5

Aerides fieldingii Jennings
(after Colonel Fielding, former British officer in the Indian army)
Fox Brush Orchid

A popular species with orchid growers which makes a highly decorative subject when well grown in a teak basket. It is native to the foothills of the Himalayas in northern India. Plants have a dense, leafy habit and pendulous racemes to 60 cm long which sometimes branch near the base. The flowers are so crowded they almost present a continuous cylinder. Each is white and waxy with purple suffusions and spots. A strong, sweet fragrance is freely released and this quickly fills a glasshouse on warm days.

Temp	Shade	In/on	Humidity	Air Flow
Hot	10-30%	Pot or Basket	30-50%	5

Aerides falcata

Aerides lawrenceae H. G. Reichb.

(after Lady Lawrence, 19th century President of the Royal
Horticultural Society)

A fine species which originates on the island of Mindanao in the
southern Philippines. It grows as an epiphyte in lightly shaded or
quite exposed positions. Plants are large (stems to 2 m long) with
thick leathery leaves and arching spikes to 60 cm long, each
crowded with large, waxy, heavily fragrant flowers which are
long-lasting and provide an impressive display. The segments are
each tipped with a prominent rich purple blotch and the labellum
is similarly marked. An easily grown species which must be given
adequate space to achieve its potential.

Temp	Shade	In/on	Humidity	Air Flow
Hot	10-30%	Pot or Basket	50-70%	5

Aerides lawrenceae

Aerides multiflora Roxb.

(with many flowers)

The natural distribution of this species extends from the
Himalayas of northern India through Burma and Thailand to Laos,
Cambodia and Vietnam. It grows on trees in bright light. Prized
for its stately growths and long, pendulous spikes of crowded,
colourful flowers, this species is extremely popular with
orchidists. Commonly its white, waxy fragrant flowers have light
purple stains and markings, but in superior clones the flowers
may be pale pink or have rich purple markings. In some
selections the flowers are very flat giving the racemes a crowded
appearance.

Temp	Shade	In/on	Humidity	Air Flow
Hot	10-30%	Pot or Basket	30-50%	5

Aerides multiflora

Aerides odorata Lour.

(sweet smelling)

This is the species upon which the genus *Aerides* is based, it being
collected by the Portugese missionary and botanist Loureiro in
Cochin China in the early 1790's. It is a very widely distributed
species extending from Southern China and Burma to Malaysia
and the Philippines. Being widespread it is also variable and
some fine clones have been selected. The best of these have large
flowers with bright purple markings, crisped labellum margins
and a yellowish spur. A popular species, this orchid is easily
grown if given plenty of heat, bright light, humidity and air
movement. Plants are often grown in pots but large flowering
plants in baskets produce an impressive display.

Temp	Shade	In/on	Humidity	Air Flow
Hot	10-30%	Pot or Basket	50-70%	5

*Aerides
odorata*

Aerides quinquevulnerum Lindley

(literally five wounds – an unusual reference to five spots on the perianth segments)

A native of the Philippines, this coarse epiphyte forms large, much-branched clumps on trees in hot, humid valleys. With its shiny leaves and arching or pendulous racemes of colourful waxy flowers, large plants can produce highly ornamental floral displays. They occupy considerable space, however, and in the tropics this species may be grown outdoors on rocks or trees in the garden.

Temp	Shade	In/on	Humidity	Air Flow
Hot-V.Hot	10-30%	Pot	30-70%	5

Aerides quinquevulnerum

AMESIELLA Schltr. ex Garay

(after Prof. Oakes Ames, founder of orchid herbarium, Harvard University)

A monotypic genus which is endemic to the Philippines. Its affinities are with *Angraceum* and in fact it was formerly included in that genus until 1972. The generic name was established by Rudolph Schlechter but was not published until 1926, one year after his death. The name was illegal because it was unaccompanied by a latin description but this situation was later corrected by Leslie Garay. *Amesiella* differs from *Angraecum* by its short, forward-pointing, shortly-hairy column and broadly rounded floral segments.

Cultivation: This orchid is fairly easily grown if given warmth, humidity and air movement. A small pot of bark, charcoal and chopped fibre is satisfactory or plants can be attached to a slab of tree fern, cork or hardwood. Growth slows in winter but plants should never be allowed to dry out completely.

Amesiella philippinensis

Amesiella philippinensis Ames

(from the Philippines)

First collected in 1906 on Mt Halcon, Luzon, by the American botanist Elmer Merrill, this species has become popular with orchidist. It grows at intermediate altitudes (500-1500 m) on trees in fairly dense rainforests. Plants are valued for their neat habit and racemes of large (4-5 cm across), white fragrant flowers. Each flower has a prominent spur which increases in length as the flower ages, eventually ending up about 10 cm long.

Temp	Shade	In/on	Humidity	Air Flow
Hot	50-70%	Pot or Slab	70-90%	5

ANGRAECUM Bory

(from the Malaysian word *angurek*, used for orchids with a growth habit like *Vanda*)

Angraecum is one of a complex group of genera from the African region, all of which have a long and prominent basal spur on the labellum. *Angraecum* is the largest genus of the group consisting of more than two hundred and twenty species. While it reaches its best development in various countries of continental Africa it also extends to the adjoining islands of Madagascar, Comoro Islands and the Mascarene Islands. Outliers from Sri Lanka, Japan and the Philippines are now placed in other genera. The more spectacular members of the genus *Angraecum* are commonly grown but the smaller-flowered botanicals are rare and are only to be found in the collections of enthusiasts. Most species have flowers in racemes but some are borne singly. The multi-flowered species are unusual in that the apical flower expands first and it is usually the largest on the raceme.

Cultivation: All species of *Angraecum* require hot, humid conditions coupled with bright light and ample air movement. The larger species are well suited to pots or slatted baskets using a coarse potting mixture whereas the straggly, semi-climbing types grow well on treefern slabs. The small species can also be grown on slabs or in small pots of fibre. Most growth occurs during the summer months and although plants slow down markedly during winter they still continue some growth. At this time of the year in temperate climates growers may hang their plants close to the glass to obtain maximum warmth and light. The large-growing species have very coarse roots and need a sufficiently large container for their unimpeded development.

Angraecum compactum Schltr.

(compact, crowded)

Although of compact growth habit this species is by no means the smallest of the genus, rather it seems to be intermediate in size between the miniatures and the large growers. Plants will eventually grow to about 30 cm tall with fairly thick, spreading leaves. Its 7 cm long flowers, borne singly or in pairs on a short raceme, are white and waxy with a broad, prominent labellum and delicate fragrance. A native of Madagascar this species grows on trees in warm humid rainforests at elevations between 700 and 1500 m.

Temp	Shade	In/on	Humidity	Air Flow
Hot	30-50%	Pot or Slab	70-90%	5

Angraecum compactum

Angraecum sesquipedale Thouars

(one and a half feet – an exaggerated reference to the length of the floral spur)
Christmas Star Orchid

In its natural state on the island of Madagascar, this orchid grows on sparse trees in low, hot, humid areas where there is plenty of air movement. Although plants may be straggly, they have prodigious root growth (roots in excess of 5 m long) and bear racemes of wonderful, white star-shaped flowers each about 16 cm across. These are waxy, long-lived and strongly fragrant but their outstanding feature is the narrow drooping spur attached to the base of the labellum and which may reach 30 cm in length. This orchid fascinated the great English naturalist Charles Darwin and he published a study of its pollination mechanism in which he made the shrewd prediction that somewhere in Madagascar was a moth with a proboscis of similar length to the labellum spur of this orchid. Forty years later and some years after Darwin's death, such a moth was discovered.

Temp	Shade	In/on	Humidity	Air Flow
Hot	10-30%	Pot or Basket	50-70%	5

Angraecum sesquipedale

Angraecum subulatum Lindley

(awl-shaped, in reference to the leaves)

A little-known slender Nigerian epiphyte that is of interest mainly to ardent enthusiasts. Plants consist of a few, pendulous, very slender stems to 60 cm long with scattered sharply pointed leaves. The small white flowers are borne singly or in pairs in the leaf axils.

Temp	Shade	In/on	Humidity	Air Flow
Int-Hot	10-30%	Pot or Slab	50-70%	5

Angraecum subulatum

ANSELLIA Lindley

(after John Ansell, English gardener who discovered this genus in tropical west Africa in 1842)

This genus of African orchids consists of two or three species, although some authorities prefer to treat it as monotypic with the solitary species being highly variable. Whatever the taxonomy these orchids are highly ornamental with colourful, fragrant flowers well displayed on terminal racemes. Robust orchids which form large clumps, they grow as epiphytes (rarely terrestrials) in positions ranging from protected to exposed in hot, humid climates. Their roots grow erect and form a white mass around the upright pseudobulbs. This is probably a litter-collecting device.

Cultivation: Species of *Ansellia* are very easy to grow if given abundant warmth, bright light and unrestricted air movement. Strong growers, they are best potted into a mixture containing particles of softwood bark, charcoal and chunks of fern fibre. Being gross feeders they respond to regular applications of liquid fertilisers and dried manure.

Plants have a dormant phase (short in *A. africana*) after flowering and should be kept on the dry side until new shoots appear.

Ansellia africana Lindley

(from Africa)
Leopard Orchid

Native to Angola, Namibia, Kenya and areas in south-western Africa, this species grows as an epiphyte from sea level to about 1400 m altitude. Plants form large clumps usually in shady forests but sometimes on fairly exposed trees and palms. A very handsome orchid, this species is valued for its large yellow flowers which are heavily spotted and blotched with dark red-brown markings. Plants have a very short dormant period and growers have noted that cultivated plants may grow nearly continuously and flower poorly. This species can be easily distinguished from the closely related *A. gigantea* by its petals being wider than the sepals and with a nearly circular labellum mid-lobe.

Temp	Shade	In/on	Humidity	Air Flow
Hot	10-30%	Pot	50-70%	5

Ansellia africana

ARPOPHYLLUM Llave & Lex.

(from the Greek *harpe*, sickle or scimitar, *phyllon*, leaf; a reference to the falcate leaves)

A small genus of two species, both being handsome orchids that are popular with growers, being valued for their long, dense, cylindrical racemes of colourful flowers. Native to Central and South America, they grow in a range of situations as terrestrials in the ground, among rocks and on trees, but always in bright light. It has been recorded on a number of occasions in the literature that these orchids grow in full sun developing into large, sturdy clumps with thick, leathery leaves.

Cultivation: Both species of *Arpophyllum* are fairly easy to grow but cultivated plants are not grown hard enough and tend to be coddled. As their habitat details indicate, they are hardy orchids which revel in bright light. Drainage must be excellent and as they have a strong root system, coarse materials are suitable. Although they can be grown on slabs these orchids are most commonly seen in pots.

Arpophyllum alpinum Lindley

(alpine)

Although similar in many respects to *A. giganteum*, this species has smaller plants and usually shorter, much sparser racemes. Also its sheathing bracts are warty and the ovary glands are brown. Plants produce an attractive floral display especially when well grown. A native of Mexico, Guatemala and Honduras, this species grows on the mountain tops in cloud forests at about 3000 m elevation.

Temp	Shade	In/on	Humidity	Air Flow
Cool	10-30%	Pot	50-70%	5

Arpophyllum alpinum

Arpophyllum giganteum Hartweg ex Lindley

(gigantic)

In its natural state this orchid grows in large masses on trees and rocks at about 1500 m altitude. It is a tough species which can withstand periods of dryness, violent storms and intense sunlight. First described from Mexico in the early 1800's it is now known from Guatemala, Honduras, El Salvador, Costa Rica, Colombia, Venezuela and Jamaica in the West Indies. Plants are dense with growth to more than one metre tall and showy cylindrical racemes about 15 cm long densely packed with colourful flowers. These are variable in colour form pale red to pink and mauve-purple. This species can be distinguished from its close relative by having smooth (not warty) sheathing bracts and black glands on the ovaries.

Temp	Shade	In/on	Humidity	Air Flow
Int	10-30%	Pot	30-50%	5

Arpophyllum giganteum

ARUNDINA Blume

(from the Latin, *arundo*, a reed; alluding to the reed-like stems)

This genus is usually treated by botanists as consisting of a single species, although that species is widespread, variable in the extreme and according to orchidists is almost certainly in need of a thorough botanical study. In all, about six species of *Arundina* have been named from different areas but modern treatments mostly lump these together. Being sun-loving orchids with an ornamental grassy habit and colourful flowers they have become popular garden plants for the tropics.

Cultivation: As garden plants in the tropics these orchids are grown in the same way as for many other ornamental perennials and are massed together in beds. They need plenty of sunshine, daily watering and respond to regular top dressing with animal manures. If grown as glasshouse plants, they must be potted into containers which are sufficiently large for the development of their strong root system.

Arundina graminifolia

Arundina graminifolia (D. Don) Hochr.

(with grass-like leaves)

Of value for its hardiness, adaptability and ever-flowering habit, this popular orchid is commonly grown as an ordinary garden plant in tropical countries. A sunny position in well-drained friable soil, abundant water and regular top dressing with rotted manures are adequate for its requirements. Plants develop into tall clumps (to 2 m tall) of slender, reed-like stems clothed with fresh green leaves. Flowers are borne in succession, each being attractively fragrant and lasting two or three days. They have some resemblance to a small *Cattleya* and are pale lilac-mauve with a darker labellum. A white-flowered form is sometimes grown. Outside the tropics this species needs to be grown in a heated glasshouse and given good light to flower.

Temp	Shade	In/on	Humidity	Air Flow
Hot	10-30%	Pot	70-90%	5

ASCOCENTRUM Schltr.

(from the Greek *ascos*, bag; *centron*, spur; in reference to the large labellum spur)

This small genus of about five species is distributed in southern China, Taiwan, Thailand, Burma, Malaysia, Indonesia and the Philippines. Vegetatively plants of this genus have growths like miniature species of *Vanda*, but these are transformed at flowering time by dazzling displays of brightly-coloured, long-lived flowers. These are flat and disposed around the racemes to present a cylindrical appearance. Colours common in the genus include brilliant pinks, reds and oranges. Species of *Ascocentrum* are very popular with orchid growers because they are so easy to culture and are free-flowering. They are widely used in breeding programs to produce interesting, colourful progeny.

Cultivation: Species of *Ascocentrum* can be grown in pots although it is more usual for them to be suspended in slatted baskets. Drainage must be excellent and common potting materials include charcoal, coarse bark and cubes of fern fibre. Plants need warmth, bright light and unimpeded air movement. Growth is active over spring and summer and slows down dramatically during winter when plants are best hung near the glass and kept on the dry side.

Ascocentrum ampullaceum

Ascocentrum ampullaceum (Lindley) Schltr.

(bottle or flask-shaped)

A dwarf species with stems usually 10-20 cm long and leathery, curved leaves 10-15 cm long, which are irregularly notched at the apex and with a prominent ridge beneath. It occurs naturally in the Himalayas of northern India and Burma, growing on trees in bright light. It is esteemed by growers for its bright rose-pink flowers. These are fairly small (about 1.4 cm across) but are long-lasting and crowded on stiffly erect racemes to 15 cm long. Plants are easily grown in a teak basket of chopped fibre or bark.

Temp	Shade	In/on	Humidity	Air Flow
Hot	30-50%	Pot or Basket	50-70%	5

Ascocentrum curvifolium (Lindley) Schltr.

(with curved leaves)

The largest member of the genus, this species is native to northern India, Burma and Thailand. It is primarily a species of the mountains and plants grow tardily and flower very poorly in lowland tropical areas. In suitable climates, however, this is a very floriferous orchid with racemes arising from every mature axil. Plants can grow to more than 1 m tall and the stiffly erect racemes are crowded with orange to reddish-brown flowers, each about 2.5 cm across. A large plant with six or eight racemes provides a colourful spectacle. The flowers last two or three weeks. Plants are easily grown and are well suited to large, slat baskets.

Temp	Shade	In/on	Humidity	Air Flow
Hot	10-30%	Pot or Basket	50-70%	5

Ascocentrum curvifolium

63

Ascocentrum miniatum

Ascocentrum miniatum (Lindley) Schltr.

(orange or cinnibar-red)

Although a dwarf orchid, with stems usually less than 10 cm long, this species flowers so prolifically that it imparts the impression of being larger growing. The brilliant orange to vermilion flowers crowd the stiffly erect racemes which grow to 14 cm tall. Each flower is only about 1.5 cm across but it has a long labellum spur and a light-coloured ovary and pedicel, all of which give the impression that the flowers are larger. The contrasting dark anther cap in the centre of each flower is an interesting feature. In nature this species grows as an epiphyte at moderate elevations, the plants tolerating considerable exposure. A very widely distributed species, it is native to northern India, Thailand, northern Malaysia and Indonesia.

Temp	Shade	In/on	Humidity	Air Flow
Hot	10-30%	Pot or Basket	30-50%	5

BARKERIA Knowles & Westc.

(after George Barker, 19th century British horticulturist)

A poorly known genus of eleven species, all from Mexico and Central America. They have been included in the genus *Epidendrum* by most authors, however, in modern treatments the genus *Barkeria* has been reinstated for this segregate group. All species grow on trees or rocks in fairly exposed situations.

Cultivation: Few of these lovely orchids have become well known in cultivation, but those tried seem adaptable and easy to grow. While they are suitable for pot culture in a moderately coarse mix of pine bark, charcoal and shredded fern fibre, most growers achieve best results on slabs of cork or treefern. They like plenty of light and good air movement. Being deciduous, these orchids have a distinct rest period after their leaves fall at which time they should be watered sparingly. Watering should be again increased with the onset of new growth and maintained at least daily until the new pseudobulbs begin to harden.

Barkeria lindleyana Bateman ex. Lindley

(after John Lindley, the father of modern orchidology)

This species was originally discovered in Costa Rica during the early 1830's by George Ure Skinner, a resident English merchant with a penchant for plants and birds. It was introduced to England in 1839 and first flowered in 1841. Valued for its elegant flowers which are about 7 cm across and ranging in colour from mauve to purple with a prominent spreading labellum, the species has proved to be adaptable and is easily grown. As well as Costa Rica, this species is native to Mexico, Guatemala and Honduras.

Temp	Shade	In/on	Humidity	Air Flow
Int	30-50%	Pot	50-70%	5

Barkeria lindleyana

BEADLEA Small

(after Chauncey Delos Beadle, 18th & 19th century American botanist)

This genus is comprised of about fifty-four species of terrestrial orchids which are distributed in the West Indies, Central America and South America. Originally most species were included in the genus *Spiranthes* but they are a segregate group and are usually classified in this separate genus.

Cultivation: Few species of *Beadlea* are encountered in cultivation but a couple have become relatively common. These species are easily grown in a well-drained fibrous potting mix. Although plants are evergreen they should be kept on the dry side over winter.

Beadlea elata (Sw.) Small ex Britton

(tall)

Originally placed in the genus *Spiranthes*, this evergreen species had thick, fleshy roots and a basal group of broad, deep green leaves each carried on a narrow petiole. A slender spike to 20 cm tall carries small, semi-nodding flowers arranged in a loose spiral. These vary in colour from brownish green to green or greenish-white with a prominent white labellum. Although mainly of interest to the enthusiast, this species adapts well to cultivation and can be propagated readily from seed. It is native to the West Indies, Mexico, Central America and parts of South America.

Temp	Shade	In/on	Humidity	Air Flow
Int	30-50%	Pot	50-70%	4

Beadlea elata

BIFRENARIA Lindley

(from the Greek, *bi*, two, twice; *frenum*, a strap; in reference to paired straps on the pollinia)

A genus of about eleven species of epiphytes and lithophytes from Brazil. They are vigorous plants which can grow into large clumps and have crowded, angular pseudobulbs and large thick leaves. The widely-opening flowers are fleshy, attractive and perfumed. A few species are fairly common in cultivation but the rest are scarcely known.

Cultivation: Species of this genus are easy to grow in a warm glasshouse, appreciating plenty of light with moderate humidity and gentle, unimpeded air movement. Watering should be heavy when plants are in active growth but is delivered sparingly after the growths have matured and during the cool temperatures and short days of winter. Some growers report shyness of flowering in these orchids.

Bifrenaria harrisoniae (J. D. Hook.) H. G. Reichb.

(after Mrs Arnold Harrison, 19th century English orchidist)

This popular orchid has been in cultivation for more than one hundred and sixty years, being first introduced to England during 1821. A native of Brazil, it grows on rocks and trees in warm, humid rainforests. It is prized for its widely-opening, waxy flowers which are usually cream to yellowish with a purple to magenta labellum. They have a strong fragrance most noticeable on warm sunny days and may last for several weeks. Plants grow best in a heated glasshouse and require fairly bright light to flower well. They are relatively easy to grow but can suffer a setback from repotting and hence should be repotted only when absolutely necessary.

Temp	Shade	In/on	Humidity	Air Flow
Int	30-50%	Pot	50-70%	4

Bifrenaria harrisoniae

Bletilla striata

BLETILLA H. G. Reichb.

(a diminutive of *Bletia*)

A small genus of about nine species restricted to China, Japan and Taiwan. All species were formerly included in the genus *Bletia* but were separated as a segregate group in 1853. They are terrestrial orchids with clustered pseudobulbs, relatively large leaves and colourful flowers. Only one species is at all common in cultivation.

Cultivation: These orchids appear to be relatively easy to grow. They have a distinct growing season followed by a dormant phase when they become deciduous. While dormant they should be kept on the dry side but when in active growth, watering should be a daily occurrence. They are suitable for culture in pots although the hardy ones can also be grown as garden plants.

Bletilla striata (Thunb.) H. G. Reichb.

(finely striped)
Chinese Ground Orchid

This orchid is still commonly sold in Australia as *Bletia hyacinthina*, an old synonym by which it was previously well known. A native of China and Japan it has been cultivated in Europe for more than one hundred and eighty years, being introduced into England in 1803. It is popular because it is one of the few hardy terrestrial orchids which grows well as a garden plant. Best results are achieved in temperate regions but plants are adaptable and will grow successfully in subtropical and highland tropical regions. Damage from severe frosts is avoided by the plants dying back to sturdy pseudobulbs during winter. New growth begins vigorously in spring with the colourful purple flowers appearing when the shoots are about half mature. An attractive white-flowered variant is known.

Temp	Shade	In/on	Humidity	Air Flow
Cool	10-30%	Pot	30-50%	5

BOLUSIELLA Schltr.

(after Dr Harry Bolus, 19th & 20th century South African botanist and orchidologist)

A segregate genus of African orchids separated from *Angraecum* by their tiny size and complex floral morphology. About six species comprise this distinctive and easily recognisable genus, all members of which are true miniatures. Plants are commonly less than 5 cm tall with fleshy, succulent leaves overlapping at the base and arranged in a fan reminiscent of an iris growth. Wiry roots spread along the branches of shrubs and trees on which these orchids perch. The tiny white flowers are arranged on opposite sides of a flattened raceme and are partly enclosed by prominent papery bracts.

Cultivation: Despite their tiny size these orchids are relatively easy to grow. With their very slender roots they are highly sensitive to waterlogging and in fact are best grown on the dry side since with their fleshy leaves they are well used to coping with short dry periods. While they can be successfully grown in a small pot, growers report that best growth is achieved on a small branch of a species such as English oak. Plants need continuous gentle, unimpeded air movement.

Bolusiella imbricata (Rolfe) Schltr.

(overlapping)

A true miniature orchid which is widespread in the warm, humid forests of tropical Africa (Zambia, Ghana, Cameroon, Uganda, Kenya). It has very shiny, succulent, fleshy leaves which store moisture for use during dry periods. Although mainly a collectors item, flowering plants of this tiny orchid have a certain charm. They are easily grown in a small pot of fine bark or on a slab of treefern, hardwood or cork.

Temp	Shade	In/on	Humidity	Air Flow
Int	30-50%	Pot or Slab	50-70%	5

Bolusiella imbricata

BRASSIA R. Br.

(after William Brass, 18th and 19th century plant collector and botanical illustrator)

A very interesting genus which is comprised of about fifty species and distributed in the West Indies, Florida, Mexico, Central America and South America. Although very closely related to *Miltonia* and *Oncidium* these orchids can be recognised by their elongated segments, distinctive floral arrangement on the racemes and the entire column and labellum. Because they are so easy to grow and rewarding with their displays of flowers, they are firm favourites with orchid fanciers and are an excellent group for the novice grower. In nature these orchids grow as epiphytes in shady locations in humid forests.

Cultivation: Mostly these orchids are grown in pots, however they can also be successful on slabs of cork or treefern. The best potting materials are treefern fibre or osmunda as some growers report that these orchids dont appreciate bark. Suitable conditions must include gentle air movement, high humidity and bright light.

Brassia verrucosa Lindley

(warty)

Easily recognised by its prominently warty labellum, this attractive species grows as an epiphyte in forests up to about 1500 m altitude in Mexico, Guatemala, Honduras and Venezuela. With its stiff arching racemes of large, spidery, pale green flowers and ease of culture this species is a firm favourite with orchid growers. The flowers are long lasting and have an attractive fragrance. The variant, commonly cultivated as *B. brachiata*, is synomous with *B. verrucosa*, the only significant difference being that it has larger flowers (sepals to about 15 cm long)

Temp	Shade	In/on	Humidity	Air Flow
Int	30-70%	Pot	50-70%	5

BRASSAVOLA R. Br.

(after Antonio Musa Brassavola, 19th century Italian nobleman and botanist)

A small genus of fifteen species which are natives of the West Indies, Central America and South America. They grow on trees or rocks in fairly exposed situations. Although the plants are compact and of modest size, their flowers are large and showy with a prominent labellum. An interesting feature shared by many of these orchids is that they are fragrant at night. See also the closely related genus. *Rhyncholaelia*.

Cultivation: Species of *Brassavola* are popular subjects for cultivation because they are generally easy of culture and well-grown plants are rarely out of flower. Most growers use small pots with small particles of bark and chopped fibre as the potting mix but many species (especially those with a cascading growth habit) grow equally well in baskets or on slabs of treefern fibre. Plants like bright light for good flowering and gently moving air.

Brassia verrucosa

Brassavola nodosa

Brassavola nodosa (L.) Lindley

(knotted or gnarled, a reference to the bulbous nodes)
Lady of the Night

A common species which occurs naturally in the West Indies, Mexico, Central America, Panama and Venezuela. A species of exposed situations in the lowlands it grows on trees and shrubs (including large cacti) from the coast (where it is commonly found on mangroves) to areas well inland and up to 500 m altitude. A firm favourite with growers because established plants are rarely out of flower. The flowers are pleasantly fragrant in the evening and vary from pure white to pale green or light yellow.

Temp	Shade	In/on	Humidity	Air Flow
Int	10-30%	Pot or Slab	50-70%	5

Brassavola tuberculata Hook.

(with knobby projections, in reference to the outer surface of the sepals)

Although *B. tuberculata* is the correct name for this orchid (published in 1829), it is still commonly grown and sold as *B. perrinii*, a name given to it by John Lindley in 1832. Of natural occurrence in Brazil and Bolivia, this species is prized by growers for its pendulous growth and abundance of greenish-yellow flowers which have a broad, white labellum. These flowers are long-lasting and attractively scented at night.

Temp	Shade	In/on	Humidity	Air Flow
Int	10-30%	Basket or Slab	50-70%	5

BROUGHTONIA R. Br.

(after Arthur Broughton, early 19th century English botanist)

A genus consisting of a single species which is native to the West Indies. Although closely related to *Epidendrum,* this genus is distinguished by having the prominent labellum spur joined to the ovary and also the column lacks a basal extension. Other species have been placed in this genus at various times but the modern view is that the genus is monotypic and these are best placed elsewhere.

Cultivation: Plants of this species resent disturbance and should not be divided or remounted except when absolutely necessary. Although pot culture can be successful plants on treefern slabs are often stronger and flower much more profusely. They should be hung where they receive bright light.

Broughtonia sanguinea (Sw.) R. Br.

(blood red)

The densely clustered pseudobulbs of this species are an unusual greyish green and while the young ones are plump, when older they become prominently furrowed. Wiry racemes to 60 cm long carry pink to bright crimson, widely opening flowers 2-3 cm across, each with a prominently veined labellum which is paler at the base. Rare variants that have been found include white-flowered and yellow-flowered plants. A native of Jamaica, this species grows on trees (occasionally large cacti) and rocks in bright light, sometimes in full sun. Plants are common on the coastal plains but also extend to 800 m altitude. The prevailing climate is strongly seasonal and plants are well adapted to long periods without rainfall. This species has been cultivated in Europe for nearly two hundred years, being first introduced to Kew in 1793. It is common for growers to overwater and excessively coddle this hardy orchid.

Temp	Shade	In/on	Humidity	Air Flow
Hot	10-30%	Slab	50-70%	5

Brassavola tuberculata

Broughtonia sanguinea

BULBOPHYLLUM Thouars

(from the Greek *bulbos*, a bulb; p*hyllon*, a leaf; a reference to the direct attachment of the leaf to the apex of the pseudobulb)

This huge genus numbers more than two thousand species and these are widely distributed in most tropical parts of the world, with a concentration in New Guinea where more than six hundred species occur. All species are epiphytes on trees or rocks. Because of the large number of species and wide distribution, it is difficult to generalise about this genus. Two segregate groups (*Cirrhopetalum* and *Megaclinium*) may be recognised by growers but they are linked by intermediate taxa and most modern botanical treatments include them with *Bulbophyllum*. Whereas those species which are highly ornamental or distinct novelties have become well known in cultivation, by the large very few species of *Bulbophyllum* are commonly grown. This is a great disappointment for many species are rewarding subjects and adapt well to cultivation. The flowers of some Bulbophyllums are of very unusual colouration and with tremulous lips or odd adornments on the perianth including hairs, cilia and unusual bits of tissue that have been described as flags. Some have obnoxious perfumes which can be quite off-putting.

Cultivation: Being such a large, diverse genus it is difficult to generalise as to cultural aspects required by these orchids. As a group, however, all species of *Bulbophyllum* respond to high humidity and abundant air movement. Most will grow happily at fairly low light levels but flowering may be enhanced with bright light. Some species grow well in cool conditions, others need heat. Those species with long-creeping rhizomes are best grown in baskets or on slabs where their growth is unimpeded. With their very fine roots the miniature species of *Bulbophyllum* are also best grown on slabs. The larger, compact species adapt well to pot culture using such materials as fern fibre, bark, charcoal and chopped sphagnum moss.

Bulbophyllum cumingii Lindley

(after Hugh Cuming, 19th century English plant collector and discoverer of the species)

Rainforests ranging from sea level up to 500 m altitude are the natural habitat of this Philippine species which is popular with growers for its attractive flowers. These are pinkish-purple and they radiate in a nearly perfectly symmetrical circle at the end of wiry flower stems. The dorsal sepal and petals are exquisitely fringed with prominent glandular hairs. Being rather sensitive to cold, this species must be given warm conditions to grow successfully.

Temp	Shade	In/on	Humidity	Air Flow
Hot	30-50%	Any	50-70%	5

Bulbophyllum cumingii

Bulbophyllum leopardinum (Wallich) Lindley
(spotted like a leopard)

Not only are the flowers of this orchid distinctly spotted (from which it derives its name) but they are also fleshy and have a prominent pink to red labellum. With its crowded pseudobulbs, thick and leathery leaves and free-flowering habit, this ornamental species is a collectors delight. Plants are very successful in a pot and will withstand more light than many species of this genus. A native of India, Burma and Thailand it grows in mountains at about 2000 m altitude.

Temp	Shade	In/on	Humidity	Air Flow
Int	10-30%	Pot	50-70%	5

Bulbophyllum leopardinum

Bulbophyllum lobbii

Bulbophyllum lobbii Lindley
(after Thomas Lobb, original collector)

A wonderful orchid valued for its large, interesting flowers which have a fragrance reminiscent of cucumbers. A native of Thailand, Indonesia and Malaysia, it commonly grows in the mountains at altitudes of 1000-5000 m. The best forms, which have flowers about 10 cm across, originate from Indonesia. This orchid is easy to grow on a slab of treefern and revels in a warm, humid glasshouse where it receives plenty of air movement.

Temp	Shade	In/on	Humidity	Air Flow
Int	50-70%	Slab or Basket	70-90%	5

Bulbophyllum medusae (Lindley) H. G. Reichb.
(the head of flowers resembles a Medusa)

A novelty orchid grown for its flowers which have thread-like segments and are crowded in a mop-like head at the end of a wiry flower stem. A native of Indonesia and Malaysia, it was first introduced into England in 1841 and has since become a collectors item. Individual flowers are small but the illusion they create massed together in the inflorescence is quite incredible. Plants are easily grown on a slab of treefern, cork or hardwood or in a shallow pot or basket filled with small particles of bark and chopped fibre.

Temp	Shade	In/on	Humidity	Air Flow
Int	50-70%	Pot or Slab	50-70%	5

Bulbophyllum medusae

Bulbophyllum putidum (Teijsm & Binnend.) J. J. Smith

(foul smelling)

This species is commonly sold in nurseries and exhibited at shows as either *Bulbophyllum fascinator* or *Cirrhopetalum fascinator*, both of which are synonyms of the above name. While the flower shape is typical of the *Cirrhopetalum* group, this species is unusual in that its flowers are borne singly. Although they have an obnoxious perfume, they are large and colourful with a very interesting ornamentation of paddles and flags which flutter with the slightest air movement. The species ocurrs naturally in northern India, Vietnam, Malaysia, the Philippines, Borneo and Indonesia.

Temp	Shade	In/on	Humidity	Air Flow
Int-Hot	30-50%	Slab	50-70%	5

Bulbophyllum putidum

CALANTHE R. Br.

(from the Greek *calos*, beautiful; *anthos*, flower; in reference to the attractive flowers)

Calanthe has a very wide global distribution being found in most parts of the world, especially the tropics. It ranges from South Africa to Japan, America and Asia. Numbering some one hundred and fifty species, most of which grow as terrestrials, it is a significant orchid genus and is popular with growers. In Japan, *Calanthe* is especially popular and some growers have extensive collections of species, variants and hybrids. The genus can be divided into two groups based on vegetative characters. The deciduous types shed their leaves annually and have large, angular pseudobulbs. The evergreen types have generally small pseudobulbs and retain their leaves for a number of years.

Cultivation: Although these orchids are terrestrials many growers use soil-less potting mixes for their growth. Common potting materials include coarse gravel, peat, fern fibre, leaf mould and perlite. The addition of some good quality loam to the mix can be useful if optimum growth is not being achieved. Those species which have a definite period of rest should be dried off as the leaves shed and watered again with the commencement of new growth. By contrast the evergreen types need to be kept moist throughout the year with an abundance of water in summer when they are growing actively.

In general these orchids are gross feeders and respond to heavy applications of fertilisers and manures during spring and summer. While many species will tolerate low light flowering is often improved in bright light. Some species will happily tolerate considerable exposure to sun. Regular gentle air movement is very important for their successful growth.

Calanthe masuca (D. Don) Lindley

(a native name)

Of natural occurrence in the Himalayas of northern India, this species grows as a terrestrial in accumulations of litter on shady, forested slopes. Plants are evergreen with strongly pleated leaves and the colourful flowers are clustered on stiff flower-stems which may reach 75 cm tall. A rewarding species for cultivation which, in Australia, is still a collectors item.

Temp	Shade	In/on	Humidity	Air Flow
Cool-Int	30-50%	Pot	50-70%	4-5

Calanthe masuca

Calanthe reflexa Maxim

(reflexed, in reference to the petals)

A hardy, dwarf species native to Japan where it grows on forested slopes and glades amid leaf litter. Plants produce a delightful display of small, colourful flowers on erect stems about 30 cm tall. Flower colour is variable and a pleasing effect is obtained when mixed colour forms are grown together. This species is quite hardy to cold with the plants becoming dormant over winter.

Temp	Shade	In/on	Humidity	Air Flow
Cool	30-50%	Pot	50-70%	4

Calanthe reflexa

Calanthe rubens Ridley

(blush-red)

This species has a similar growth habit to *C. vestita* and the two also share many other features. *C. rubens* however, is smaller in most of its parts and has much more colourful flowers, these being blush pink with a prominent red central stripe on the labellum. Its flowering habit is interesting for the raceme elongates as the flowers open, and flowering proceeds over a long period with the racemes finally reaching about 50 cm in length. The species is native to Malaysia and Thailand where it grows on limestone formations. Its cultural requirements are similar to those of *C. vestita*.

Temp	Shade	In/on	Humidity	Air Flow
Int	30-50%	Pot	50-70%	5

Calanthe vestita Lindley

(clothed)

A deciduous species which grows as a terrestrial in Thailand, Burma, Vietnam, Malaysia and Indonesia. It grows in rocky areas (often limestone) which have a strong seasonal climate, the plants shedding their leaves in the dry season and growing strongly with the advent of good rains. Cultivated plants rot readily if kept too wet while dormant. A good policy is to keep dormant plants completely dry until new shoots appear and then pot into new mixture and commence regular watering. The flowers of this species are beautiful and long-lasting. They are variable in colour, some being wholly white with a yellow labellum callus, others with a prominent red blotch on the labellum and some with the whole labellum red. This popular species requires more light than the evergreen species of *Calanthe* but still needs protection from direct sun.

Temp	Shade	In/on	Humidity	Air Flow
Int	30-50%	Pot	50-70%	5

Calanthe rubens

Calanthe vestita

CATASETUM Rich.

(from the Greek, *cata*, downwards; *seta*, bristle; in reference to the downcurved, antennae-like appendages attached to the base of the column of male flowers)

Interesting, bizarre, fascinating and beautiful, all words which could be applied to this astonishing genus of orchids. Numbering some one hundred and ten species, the genus is found in Central America with a predominance in Brazil. Most species are epiphytes but some grow on rock outcrops and an occasional few are found as terrestrials in sandy soils. All species are deciduous and they are found from sea level to an altitude of about 1500 m. The bulk of the species grow in very bright light in sparse forests often on the trunks of palms. Those species from the higher mountains, however, usually grow in shady forests. Those species originating in the Amazon region have nearly a year-round growth cycle whereas the species from areas which have a distinctly seasonal climate have definite dormant and active growth phases.

An interesting feature of species of *Catasetum* is that they bear separate male and female flowers and each flower type is very different. In fact they are so distinctive that early botanists classified the separate sexes as different species. Plants growing in bright sun commonly produce female flowers whereas male flowers usually only occur on plants growing in shade. On an individual plant, the different sexes are commonly produced in different seasons, but occasionally both appear at the same time. Male flowers are well-spaced on long, arching, slender inflorescences whereas female flowers are clustered together on a thicker inflorescence. Rarely do flowers of both sexes occur on the same inflorescence.

Cultivation: Species of *Catasetum* are generally easy to grow in pots or baskets in mixes containing chopped fibre, fine particles of bark and charcoal together with some sphagnum moss. They require very bright light and abundant air movement. Growth begins in late spring and early summer with the pseudobulbs maturing over summer and flowering in autumn and winter. After flowering the plants lose their leaves and become dormant. While dormant, watering should be withheld until the onset of new growth. Plants are capable of extremely rapid growth and during the growing cycle they require heat, moisture and plenty of nutrients. Large, well-grown pseudobulbs carry up to four racemes each season.

Catasetum macrocarpum Rich.

(with large fruit)

Typically an orchid of equatorial zones of the Amazon region, this species also extends to coastal districts near Rio de Janeiro where night temperatures in winter dip to about 10°C. A vigorous epiphyte, plants of this species can grow to 45 cm tall and have arching flower spikes of similar length. The fleshy flowers have a strong, unpleasant odour but nevertheless this is an interesting orchid to grow. Plants originating from the Amazon region have a very short dormant period whereas those from coastal districts are dormant for two or three months.

Temp	Shade	In/on	Humidity	Air Flow
Hot	30-50%	Any	50-70%	5

Catasetum macrocarpum

Catasetum oerstedii H. G. Reichb.

(derivation unknown)

Cultivated plants of this species often only produce male flowers; these arise on separate inflorescences from the females. The male flowers are strongly hooded and borne up to twelve at a time on long, arching racemes. By contrast the female flowers are not hooded and two or three are carried on stiffly erect racemes. The male flowers have a heavy, sweet perfume and last well. Native to Nicaragua, Costa Rica, Panama and Colombia, this robust species grows in humid, lowland forests.

Temp	Shade	In/on	Humidity	Air Flow
Hot	30-50%	Pot	50-70%	5

Catasetum oerstedii

Catasetum pileatum H. G. Reichb.

(with a cap like the head of a mushroom)

One of the more popular members of the genus with orchid growers, *C. pileatum* is valued principally for the large, showy male flowers. These open widely, have a waxy lustre, emit a delightful fragrance and occur in a wide variety of bright hues from white to lime green and rich butter yellow. This species is the National flower of Venezuela and is also found in Brazil, Colombia, Ecuador and Trinidad in the West Indies.

Temp	Shade	In/on	Humidity	Air Flow
Hot	30-50%	Pot	50-70%	5

Catasetum pileatum

CATTLEYA Lindley

(after William Cattley, 19th century English horticulturist and orchid enthusiast)

This genus, perhaps more than any other, has captured the enthusiasm and imagination of orchid growers throughout the world. Many growers specialise in the cultivation of Cattleyas and related genera, with flamboyant colourful hybrids largely replacing the species in many collections. In all there are about sixty-five species of *Cattleya*, all being restricted to Central America and South America. They grow on trees or rocks often in very exposed situations and range from the lowlands to above 2,500 m altitude. In parts of their range the climate is strongly seasonal with distinct wet and dry seasons. This seasonality is reflected in the distinct dormant and active growth stages of many species. Within the genus *Cattleya* two growth habits can be recognised. The bifoliate group typically have two leaves, long cylindrical pseudobulbs and relatively small flowers in multi-flowered racemes. In the second group, termed the unifoliate Cattleyas or 'Labiatae' group, the stocky pseudobulbs have a single leaf and the spectacular flowers are larger and carried about one to four per raceme.

Cultivation: Members of this genus are generally regarded as being easy to grow, however a few species also have the reputation of being difficult with the frequent result being a slow decline in health and vigour. While species of *Cattleya* can be grown successfully on slabs, most orchidists grow them in pots, less commonly in baskets. Potting materials include fern fibre (particularly osmunda), softwood bark, charcoal and perlite. Allow sufficient room in the pot for two years growth and pot all plants tightly. Watering should be daily as long as warm conditions prevail and active root growth is present. As weather cools or the root growth tapers off, so the watering is reduced in frequency. Growers commonly recommend about 15°C as a minimum temperature for species of *Cattleya*, however some of those from higher altitudes will tolerate lower temperatures. Humidity shoud be high with continual gentle air movement.

Cattleya aclandiae Lindley

(after Lady Acland of Exeter, England who first introduced the species in 1839)

A native of the northern province (called Bahia) of Brazil, this species grows on scattered trees in sparse lowland forests often close to the coast. Plants grow in fairly exposed conditions with almost constant air movement. A fairly small-growing species which is prized for its fleshy, long-lasting flowers of a most unusual colouration. Typically the sepals and petals are yellowish-green heavily barred with purplish-black and the protruding labellum is bright magenta-purple. This species is easy to grow if given warmth, abundant air movement and bright light. This species is spelt by some authorities as *C. acklandiae*.

Temp	Shade	In/on	Humidity	Air Flow
Hot	10-30%	Pot, Basket, Slab	50-70%	5

Cattleya aclandiae

Cattleya amethystoglossa Lindley & H. G. Reichb.

(with an amethyst or bluish-mauve lip)

This species has had a confused botanical history and is still treated by some authorities as a form or variety (*Prinzii*) of *Cattleya guttata*. It is also sometimes linked with *C. porphyroglossa*. It is distinguished from both primarily on flower colour, arrangement of the floral parts and the prominent spots on the sepals and petals. Plants also often have larger flowers more numerous in the racemes. Whatever its status this orchid is certainly attractive with its waxy pink to pale purple flowers spotted heavily with amethyst and with a prominent bluish-mauve labellum which spreads broadly on the mid-lobe. A native of northern Brazil, it grows on isolated trees in sparse forests of the lowlands, often in intense sunlight with the plants taking on heavy purple-red pigmentation.

Temp	Shade	In/on	Humidity	Air Flow
Hot	10-30%	Pot, Basket or Slab	30-50%	5

Cattleya amethystoglossa

Cattleya bowringiana Veitch

(after J. C. Bowring, 19th century English orchid grower)

This orchid is popular because of its highly attractive floral displays, with up to fifteen colourful flowers being crowded into each inflorescence. Large specimen plants produce a most impressive appearance. Of natural occurrence in British Honduras and Guatemala, this orchid grows on rocks and cliffs in ravines and gorges along streams, often in full sun. A number of excellent selections based on flower colour have been made, with some having dark purple flowers. Plants of this species are easy to grow and lend themselves well to specimen culture.

Temp	Shade	In/on	Humidity	Air Flow
Int	10-30%	Pot	50-70%	5

Cattleya dormaniana

Cattleya dormaniana H. G. Reichb.

(after Charles Dorman, 19th century English grower in whose collection this species first flowered)

A lovely species which is reputedly difficult to maintain in cultivation, however consideration of its habitat can make success achievable. a native of Brazil, this orchid occurs on rocks and trees in very high rainfall tropical forests and grows on eastern and southern slopes between 400 and 1000 m elevation. Cultivated plants do not like to dry out and they are also very sensitive to salt build-up from excess fertiliser use. Hence the drainage of the potting mixture must be completely unimpeded. This species will also grow and flower well with less light than will most members of this genus.

Temp	Shade	In/on	Humidity	Air Flow
Int-Hot	30-50%	Pot	50-70%	5

Cattleya forbesii Lindley

(after M. Forbes, original collector)

Once common in the vicinity of Rio de Janeiro, this Brazilian species is now regarded as being rare in its native state. It has been recorded as growing on low trees and rocks near the sea, the prevailing climate being strongly seasonal. Plants were first introduced to England in 1823 but this species is less showy than most and is mainly favoured by collectors. Plants are generally easy to grow and flower regularly each year.

Temp	Shade	In/on	Humidity	Air Flow
Hot	10-30%	Pot or Slab	50-70%	5

Cattleya forbesii

Cattleya bowringiana

Cattleya guttata

Cattleya guttata Lindley

(spotted, dotted)

This, the largest growing species of *Cattleya*, is still common today in its native country of Brazil. Plants grow in a wide range of habitats from low scrub near the sea to rocks and taller trees further inland, but always in very bright light, often even in full sun. The prevailing climate is strongly seasonal with a torrential summer wet season followed by a long, hot, dry winter. *C. guttata* adapts well to cultivation and with the colour variations available, is an excellent species for enthusiasts to grow.

Temp	Shade	In/on	Humidity	Air Flow
Hot	10-30%	Pot	30-50%	5

Cattleya harrisoniana Batem. ex Lindley

(after Mrs Harrison, wife of the nurseryman who introduced it to England)

Although often treated as a variety of *C. loddigesii* this species is taller growing with very flat flowers which have a prominent deep yellow or orange callus on the labellum which in turn has its lateral margins reflexed. A native of Brazil it grows in bright light on trees in sparse forests.

Temp	Shade	In/on	Humidity	Air Flow
Hot	10-30%	Pot or Slab	50-70%	5

Cattleya labiata Lindley

(with a lip – in reference to the prominent labellum)

This species gives its name to a group of Cattleyas known as the Labiata group. Some taxonomists treat the numerous variants found in this group as varieties, others prefer to treat them at specific rank. Whatever their botanical status they include some very beautiful and often spectacular orchids. All members of the group have a single leaf on each mature pseudobulb and large colourful flowers which have a prominently flared, frilly labellum. Typical *C. labiata* is of Brazilian origin but variants of the group are widespread in Central America and South America. *C. labiata* (in the narrow sense) is believed to be extinct in the wild.

Temp	Shade	In/on	Humidity	Air Flow
Hot	10-30%	Pot	50-70%	5

Cattleya labiata

Cattleya loddigesii Lindley

(after Conrad Loddiges, founder of a famous English orchid nursery)

A native of Brazil and Paraguay, this species had achieved immense popularity and is firmly entrenched in cultivation. It was first introduced to Europe during the early 1800's by the Loddiges nursery, after whom it is named. In nature it grows on trees and rocks from the lowlands to the mountains, in situations ranging from shade to nearly full exposure. Plants have a neat growth habit and flower freely.

Cattleya harrisoniana

Cattleya loddigesii

Caularthron bicornutum

Cattleya velutina H. G. Reichb.

(velvety, covered with short, soft hairs)

When this species was first found there was much speculation that it was a natural hybrid between *C. bicolor* and a variant of *C. guttata*. It has been shown to be common in some areas however and is accepted as a species in its own right. Its flowers are unusually coloured but are valued for a very pleasant, pervading fragrance reminiscent of violets. They are borne in clusters of up to seven in a raceme and are glossy, fleshy and long-lasting. A native of Brazil, this species is easy to grow.

Temp	Shade	In/on	Humidity	Air Flow
Hot	10-30%	Pot or Slab	50-70%	5

CAULARTHRON Raf.

(from the Greek, *caulos*, stem; *arthron*, joint; the persistent leaf bases impart the impression that the elongated pseudobulbs are jointed)

Members of this small genus, which consists of about six species, are commonly grown under the wrong name of *Diacrium*. Native to Central America, South America and Trinidad, they grow on trees and rocks in fairly bright light. Their pseudobulbs are often hollow and plants may be fiercely guarded by nests of ants which can make collection difficult. A couple of species of *Caularthron* are popular with growers because they are attractive and free flowering.

Cultivation: In general, orchids of this genus are easy to grow and rewarding but are sensitive to disturbance and should be divided infrequently. They require bright light, warmth, humidity and air movement. Plants become dormant over winter and grow strongly in spring and summer when they should be watered and fertilised heavily.

Caularthron bicornutum (Hook.) Raf.

(with two horns, in reference to the prominent pair of calli on the labellum)
Virgin Orchid

A very attractive orchid with waxy, white, fragrant, long-lived flowers to 7 cm across which are carried on a wiry raceme to 40 cm long. It has been described as among the finest of American orchids and is popular with growers. In nature plants grow in coastal districts, in warm, humid conditions with bright light.

Temp	Shade	In/on	Humidity	Air Flow
Hot	10-30%	Pot	50-70%	5

Cattleya velutina

CERATOSTYLIS Blume

(from the Greek, *cerato*, horn; *stylis*, style; in reference to the fleshy, horn-like column)

A genus of about seventy species distributed from northern India through Malaysia to Indonesia, New Caledonia and New Guinea where about half of the species occur. They are unusual orchids with prominent sheaths covering the stems and hiding the point of attachment of the leaves. Two segregate groups have been established based on stem length. Those with short stems have a crowded habit whereas in the long-stemmed group they are sparse and often pendulous. All species are epiphytes, with many growing in the upper canopies of trees in dense rainforest. Most are small-flowered and are mainly of botanical interest.

Cultivation: Those species with a creeping or pendulous growth habit are generally unsuited to growing in pots and are best established on slabs of cork or treefern. The compact growers are easily contained in a small pot. Because of their very fine roots these orchids demand excellent drainage. Air movement should be gentle and unimpeded and humidity high.

Ceratostylis retisquamea

Ceratostylis retisquamea H. G. Reichb.

(with netted scales, in reference to the prominent bracts which cover the stems)

Previously well known as *C. rubra*, this attractive species is native to the Philippines where it grows on the islands of Luzon and Mindanao, on trees in rainforest. Plants have a neat habit of growth with the stems being clothed by large brown papery bracts and contrasting with the fleshy, dull green prominently grooved leaves. The surprisingly large widely opening flowers (about 2.5 cm across) are bright red and shiny. On well-grown plants they are produced in abundance, hiding the bracts and producing a most impressive display. This species grows well on a slab of treefern or cork.

Temp	Shade	In/on	Humidity	Air Flow
Hot	50-70%	Slab	70-90%	4

CHILOSCHISTA Lindley

(from the Greek *cheilos*, a lip; *schistos*, split or cleft; in reference to the cleft or notched labellum)

Members of the genus *Chiloschista*, which consists of about four species, are all small leafless epiphytes which have prominent roots that may be flat or round in cross section. The roots contain chlorophyll and can function in the same way as leaves. The central stem of these orchids is much reduced and covered with bracts and the small flowers are generally short lived.

Cultivation: Being leafless these species can be sensitive to disturbance, but once established on a slab they seem to grow quite readily. A slab or weathered hardwood is usually successful. It should be hung in bright light and the plants given plenty of warmth, humidity and air movement.

Chiloschista lunifera

Chiloschista lunifera (H. G. Reichb.) J. J. Smith
(bearing small moon-shaped structures)

A native of northern India and Burma, this species grows as an epiphyte in warm, humid lowland forests. Plants form a tangled mass of somewhat zig-zagged roots and have wiry racemes to 15 cm long which bear attractive, widely opening yellow flowers heavily marked and spotted with red-brown. Although relatively easy to grow, this species is mainly found in the collection of ardent enthusiasts.

Temp	Shade	In/on	Humidity	Air Flow
Hot	10-30%	Slab	50-70%	5

CHYSIS Lindley
(from the Greek, *chysis*, melting; a reference to the apparently fused pollinia in the self-pollinating species, *C. aurea*)

A small genus consisting of two or four species depending upon which botanical authority is followed. Natives of Mexico, Central America and South America, these orchids grow on trees or rocks in conditions of bright light. Plants have a distinct dormant period followed by rapid new growth. New roots are produced from the base of the new shoots, the latter being initially leafy but as the pseudobulbs mature the leaves are shed. Flower spikes arise with the new growths.

Cultivation: These orchids are capable of vigorous growth and usually have strong roots. With their pendulous pseudobulbs, plants are best grown in a basket or hanging pot. They require excellent drainage and need to be potted into a coarse mixture made from coarse pieces of terracotta, bark, charcoal and fern cubes. While in active growth, the plants should be watered heavily and fertilised at least once a week. When the pseudobulbs have matured, plants need to be kept dry and cooler until new growth commences again.

Chysis bractescens Lindley
(with prominent bracts)

Some authorities consider this to be a variety of *C. aurea* Lindley, however, it can be distinguished from that species by the prominent large bracts, smaller ovaries and the white flowers which are not self pollinating. In nature it grows at altitudes of up to 800 m altitude in humid forests, usually on trees but sometimes on calcareous rocks. A native of Mexico and Central America this orchid is popular in cultivation for its large, waxy, fragrant, long-lasting flowers. Plants grow readily if promoted by heavy watering, fertilising and warmth over summer and kept dry over winter.

Temp	Shade	In/on	Humidity	Air Flow
Hot	10-30%	Basket	50-70%	5

Chysis bractescens

CLOWESIA Lindley
(after Rev. Clowes, a keen English orchid grower who first flowered a species of the genus)

A small genus of about five species which are found in Mexico, Central America and South America. A distinctive segregate group related to *Catasetum*, they can be distinguished by having bisexual flowers, the labellum being firmly fixed to the base of the column and the pollinia being released by lifting the anther cap (not triggered by pressure as in *Catasetum*). These very attractive orchids grow as epiphytes and have a distinct growth and dormancy cycle. They become deciduous in winter after rapid growth over summer.

Cultivation: As for species of *Catasetum*. Plants must be kept dry over winter until growth commences in spring.

Clowesia russelliana (Hook,) Dodson
(after John Russell, sixth Duke of Bedford, a keen gardener and promoter of botany)

The large green flowers of this orchid are borne in attractive arching racemes. They are strongly fragrant and readily identified by the prominently humped, deep spur on the labellum. Plants flower as the new growths mature or even after the leaves have fallen. A native of southern Mexico and Central America south to Nicaragua, this orchid grows on trees and rocks in situations of bright light.

Temp	Shade	In/on	Humidity	Air Flow
Hot	30-50%	Pot, Basket or Slab	50-70%	5

Clowesia russelliana

COELOGYNE Lindley

(from the Greek, *coelos*, hollow; *gyne*, woman; apparently alluding to the deep stigmas)

This large genus of about two hundred species of orchids, has its centre of development in India, but is widely dispersed over a huge area including China, South-east Asia, Malaysia, the Philippines, Indonesia, New Guinea and Fiji. The orchids grow on trees or rocks often in bright but diffuse light. They are known from lowland jungles which are hot, wet and humid for most of the year and extend to the mountains above 2000 m where the climate is strongly seasonal and plants only produce growth during the wet summers. As a group, *Coelogyne* is very distinctive both in vegetative and floral features.

Cultivation: Popular in cultivation, species of *Coelogyne* are generally easy to grow and flower freely. All species must have excellent drainage, bright but diffuse light, high humidity and unimpeded air movement. Those species from the tropical lowlands require year round warmth and high humidity, whereas those from the mountains are quite cold tolerant and have a distinct period of rest when water should be applied sparingly. As a group these orchids are well suited to specimen culture and in fact most species suffer a setback when repotted and are best disturbed as little as possible. Those species which have long pendulous racemes make admirable basket plants.

Coelogyne cristata Lindley

(crested)

The type of the genus, this species was discovered in northern India in 1824 by Nathaniel Wallich and named by John Lindley the following year. A long time favourite with growers in Australia because of its ease of culture and free-flowering habit, this species is quite cold tolerant and grows well in temperate regions. Its attractive, large, white flowers have a pleasant fragrance and are carried on slender, arching racemes to 30 cm long. Plants have chubby, shiny, crowded pseudobulbs and the dark green, often wavy, leaves are very distinctive. An excellent species for a basket, but can also be grown in a pot or on a slab, which lends itself well to development as a specimen.

Temp	Shade	In/on	Humidity	Air Flow
Cool	30-50%	Pot or Slab	50-70%	5

Coelogyne cristata

Coelogyne flaccida Lindley

(limp, withered, flabby)

A useful basket plant, this species has a compact growth habit and pendulous racemes to 30 cm long which can carry up to twelve flowers. Large, well-grown plants can produce an impressive floral display. The white flowers, which present themselves well, are strongly fragrant, so much so as to be regarded by some growers as overpowering. A native of northern India where it extends to fairly high elevations in the Himalayas, this species grows very well in warm temperate and subtropical regions.

Temp	Shade	In/on	Humidity	Air Flow
Cool-Int	30-50%	Pot, Basket or Slab	50-70%	5

Coelogyne flaccida

Coelogyne fuliginosa Hook.

(brownish-black, sooty)

A native of southern China, Hong Kong, northern India, Thailand and Vietnam, this species grows in the foothills at moderate elevations (to about 1200 m) on exposed rocks, less commonly on trees. The flowers are somewhat variable in colour depending on their origin, some being basically greenish while others are pale yellow through to light brown. The labellum is the most attractive feature with its dense fringing on the margins and the crisped, dark brown calluses and radiating lines. With its long-creeping rhizomes, this species is best grown on slabs or in a basket.

Temp	Shade	In/on	Humidity	Air Flow
Int	30-50%	Basket or Slab	50-70%	5

Coelogyne fuliginosa

Coelogyne massangeana H. G. Reichb.

(after M. Massange de Louvrex, a 19th century Belgian orchid enthusiast)

One of the most popular and successful orchids for culture in baskets and hanging pots, this species produces strongly pendulous racemes to 60 cm in length which carry up to twenty flowers evenly spread like a necklace. The racemes are capable of extremely rapid growth and all of the flowers open within a short time of each other. Pale yellow, about 6 cm across and strongly fragrant, the flowers have a prominent brown labellum with a distinctly warty surface. The species is native to Indonesia, Malaysia and Thailand growing on ridges and slopes in the mountains.

Temp	Shade	In/on	Humidity	Air Flow
Int	30-50%	Basket	50-70%	5

Coelogyne massangeana

Coelogyne mayeriana H. G. Reichb.

(derivation unknown)

A native of Indonesia and Malaysia, this species grows in humid lowland forests, often close to the sea. Plants have a wide-creeping rhizome with well-spaced pseudobulbs, and hence are best grown in baskets or on slabs where they have room to spread. Arching or pendulous racemes carry lovely green flowers, each about 7 cm across, with a prominent, frilly labellum having black markings and white warts near the end. These warts and the broadly flared labellum mid-lobe serve as a ready means of separating this species from the closely related C. pandurata.

Temp	Shade	In/on	Humidity	Air Flow
Hot	30-50%	Slab or Basket	70-90%	5

Coelogyne mayeriana

Coelogyne nervosa A. Rich.

(having veins, often referring to prominent veins, probably in the leaves)

This species is commonly grown in Australia as C. corrugata, but the earliest and correct name is C. nervosa. A hardy species which is endemic in southern India where it grows on rocks, often in full sun. Plants can be recognised easily by the pale yellow, wrinkled pseudobulbs and broad, leathery leaves with prominent veins. Three or four large white flowers are carried on each semi-drooping raceme and they have prominent reddish or mustard-coloured markings on the labellum. A rewarding and easy species to grow.

Temp	Shade	In/on	Humidity	Air Flow
Hot	10-30%	Pot or Basket	50-70%	5

Coelogyne nervosa

Coelogyne pandurata Lindley

(fiddle-shaped)

The beautiful flowers of this orchid always excite attention and even experienced growers anticipate with pleasure its annual flowering. Each flower, which is about 10 cm across, has bright green sepals and petals and a yellowish labellum strikingly veined and blotched with black. These flowers are borne up to six at a time on arching racemes and present themselves in a most decorative manner. The plants themselves are vigorous with leathery leaves to 60 cm long and widely spaced, compressed pseudobulbs. A native to the steamy rainforests of Sumatra, Borneo and Malaysia this species requires hot humid conditions throughout the year. Best growth is achieved in hanging containers and some growers suspend the plants over containers of water to maintain continuous high humidity. This orchid should be repotted with care, preferably at the onset of new growth.

Temp	Shade	In/on	Humidity	Air Flow
Hot	50-70%	Pot	70-90%	5

CYCNOCHES Lindley

(from the Greek, *cycnos*, swan; *auchen*, neck; in reference to the slender, orchid, swan-neck-like column of the male flowers)

Commonly known as 'Swan Orchids', a derivation no doubt arising from their generic name, these fascinating orchids are prized for their handsome flowers. In all there are about twelve species in the genus but some botanical authorities reduce this to seven, treating many as mere varieties. The genus is found in Mexico, Central America and South America, the species growing as epiphytes in conditions of bright light. All are deciduous with the leaves being shed after an extremely active growing period and preceding a dormant phase.

Like the genus *Catasetum* these orchids have separate male and female flowers, the males being easily recognised by the long, very slender column (that of the females is short and stout). Each flower type may be either alike or very different in appearance. In fact the sexes of some species have flowers so dissimilar that they were described as separate species by early botanists. In addition within a species, the male flowers are often extremely variable in shape and structure, and this makes accurate identification of some species extremely difficult. Male and female flowers may be mixed together on the same raceme but are more usually found on separate inflorescences.

Cultivation: Species of *Cycnoches* require an abundance of light, warmth, water and fertiliser while they are in active growth over summer. After flowering the plants lose their leaves and become dormant. During this phase watering should be withheld until the production of new shoots in spring. While in their growing cycle, plants are capable of extremely rapid growth and everything should be done to ensure large, plump growths are produced. Mostly these orchids are grown in baskets containing chopped fern fibre, fine particles of bark and charcoal and some sphagnum moss.

Cycnoches warscewiczii H. G. Reichb.

(after Joseph Ritter von Rawicz Warscewicz, 19th century Polish collector active in South America)

This species is treated as a variety of *C. egertonianum* by some authors but it can be readily distinguished by its short erect racemes of flowers which have an entire labellum. It grows on trees, rotting logs or less commonly in the ground in humid forests up to 1000 m altitude in Guatemala, Costa Rica and Panama. Prized for its greenish-yellow flowers which can be more than 14 cm across, this species grows quite readily in cultivation but is very sensitive to poor drainage. In this species of *Cycnoches* the male and female flowers are similar but the female flowers are larger and have a much thicker column.

Temp	Shade	In/on	Humidity	Air Flow
Hot	10-30%	Pot or Basket	50-70%	5

Coelogyne pandurata

Cycnoches warscewiczii

CYMBIDIUM Sw.

(from the Greek *cymbe*, a boat or boat-shaped; *idium* a diminutive suffix; a reference to a small, hollow, boat-shaped recess found on the labellum of some species)

The most recent assessment of this genus shows that it consists of forty four species, a number of varieties and numerous minor variants which are probably best treated as cultivars. Distributed from northern India and China to Australia, these orchids grow as epiphytes, lithophytes and, less commonly, terrestrials. Some species are common in humid, lowland forests (often growing in bright light) whereas others are found in the mountains above 1800 m altitude. Many of the species forming large clumps with prominent thickened pseudobulbs but a few species have long slender stems and do not form pseudobulbs. One species is a leafless saprophyte.

More than any other orchid genus, *Cymbidium* has been subjected to an intensive selection and breeding program which has resulted in the production and registration of thousands of hybrids. Valued for their ease of culture and prolific production of large, colourful flowers, these orchids are the basis of cut flower industries in many temperate countries. More recently deliberate crossings have reduced the flower size while still retaining colour and these cultivars are sold as attractive flowering pot plants. Small-flowered hybrids having pendulous inflorescences are unexcelled as basket plants.

Cultivation: Basically, species of *Cymbidium* are easy orchids to grow, although the volume of written cultural advice would fill tomes. Mixtures must drain very freely and can be composed of materials such as bark, chopped fern fibre, charcoal, limestone chips and terracotta chards. A good potting mixture should be able to last for two or three years after which repotting will be necessary. Cymbidiums like plenty of water when growing but should be kept just moist after the pseudobulbs mature and prior to the production of flower spikes. Aeration should be free and unimpeded and as a general rule these orchids require bright light, some even being happy in full sun.

Cymbidium elegans Lindley

(graceful, neat, elegant)

Once common in orchid collections in southern Australia, this attractive species is now rarely seen. It forms a neat pot plant and the elegant, arching racemes of crowded, bell-shaped, cream or pale-yellow flowers are most attractive. It is common in northern India, northern Burma and south-western China, typically growing on shaded trees or rocks in mountainous regions with altitudes between 1400 and 2500 m.

Temp	Shade	In/on	Humidity	Air Flow
Cool	30-50%	Pot	50-70%	5

Cymbidium elegans

Cymbidium erythrostylum Rolfe

(with a red style)

This *Cymbidium* is unusual in that the flower spike arises early with the new growth and the petals and lateral lobes of the labellum cover the column. Details of this species in its natural state in Vietnam are sketchy but it apparently grows as an epiphyte in ranges at about 1500 m. Prized by growers for its early flowering habit and large glistening white flowers with a strikingly marked labellum, it is a rewarding subject which takes well to cultivation.

Temp	Shade	In/on	Humidity	Air Flow
Int	30-50%	Pot	50-70%	5

Cymbidium erythrostylum

Cymbidium floribundum Lindley

(with many flowers)

This commonly grown species is better known by the familiar, but incorrect name of *C. pumilum*. Of natural occurrence in southern China and Taiwan, this species grows on rocks and trees in situations ranging from shaded gorges to full sun at altitudes between 800 and 2800 m. The prevailing climate is strongly seasonal and plants are able to withstand several weeks of dryness. This orchid has recently become naturalised in the warmer parts of Japan. Prized for its compact growth and pendulous racemes of colourful flowers, this species is unexcelled as a specimen for basket culture.

Temp	Shade	In/on	Humidity	Air Flow
Int	30-50%	Basket	30-50%	5

Cymbidium floribundum

Cymbidium iridioides D. Don

(resembling iris)

Previously well known as *C. giganteum*, this species is of natural occurrence in northern India, Bhutan, Burma and south-western China. It is reported to grow on mossy trees in shady, humid forests between 1200 and 2200 m with the most vigorous plants gaining foothold in hollows. This species is frequently confused with *C. tracyanum* but has smaller flowers with very short hairs on the labellum and normal spreading petals.

Temp	Shade	In/on	Humidity	Air Flow
Cool	30-50%	Pot	50-70%	5

Cymbidium iridioides

Cymbidium lancifolium

Cymbidium lancifolium Hook.

(with lance-like leaves)

A *Cymbidium* with many unusual features, this species grows as a terrestrial in leaf litter in shady, humid forests often on ridges and slopes. Its distinctive, cigar-shaped pseudobulbs are held above the substrate by strong, stilt-like roots and the broad leaves are unique in the genus. Although small, the white to pale green, occasionally striped and spotted flowers are attractive and are borne on erect racemes which emerge from about half way up the pseudobulb. Easily the most widely distributed member of the genus, this species extends from northern India to New Guinea.

Temp	Shade	In/on	Humidity	Air Flow
Cool-Int	50-70%	Pot	70-90%	4

Cymbidium mastersii

Cymbidium mastersii Griffith ex Lindley

(after Dr Masters, former superintendent, Calcutta Botanic Gardens)

Native to northern parts of India, Thailand and Burma, this species grows on trees, rocks and rotting logs, at high altitudes (to 1800 m) usually in deep shade. Plants lack distinctive pseudobulbs and new leaves are produced indeterminately from the apex so that old stems become quite elongated. White, fleshy almond-scented flowers, each about 6 cm across, are borne on stiff racemes which may be erect and arching or pendulous. Although very attractive, this species is uncommonly grown and is found only in the collections of enthusiasts.

Temp	Shade	In/on	Humidity	Air Flow
Cool	50-70%	Pot	50-70%	5

Cymbidium parishii

Cymbidium parishii H. G. Reichb.

(after the Rev. Charles S Parish, original collector)

A beautiful species which is native to northern Burma where it grows in the mountains at elevations of about 1800 m. Apparently rare in cultivation, it is often confused with *C. eburneum* but the two are quite distinct. Both species flower from the upper axils of the pseudobulb which continues elongating for a few years and both have unequally notched leaf tips. *C. parishii* however has larger flowers than *C. eburneum* and is readily distinguished by its boldly marked labellum.

Temp	Shade	In/on	Humidity	Air Flow
Cool-Int	30-50%	Pot	50-70%	5

Cymbidium traceyanum L. Castle

(after A. H. Tracey, who purchased the original plant as *C. lowianum*)

One of the earliest flowering species of *Cymbidium*, this orchid is popular for its ease of culture and delightfully scented flowers. Of natural occurrence in China, northern Burma and northern Thailand, plants grow on trees in dappled shade at elevations of between 1200 and 1900 m. Plants characteristically have erect aerial roots which are probably an adaptation for trapping litter. Although confused with other species, *C. tracyanum* can be readily distinguished by its prominently hairy labellum and strongly curved petals.

Temp	Shade	In/on	Humidity	Air Flow
Cool	10-30%	Pot	30-50%	5

Cymbidium traceyanum

DENDROBIUM Sw.

(from the Greek, *dendros*, tree; *bios* life; in reference to the epiphytic habit of most species)

As it stands at present, this genus is made up of more than 1600 species and new discoveries are being made each year. *Dendrobium*, however, consists of a number of distinct segregate groups and there are conflicting views as to whether these should be recognised as sections within a large genus or separate genera. The group is widely distributed from India, Southeast Asia and Japan through Malaysia to the Philippines, Indonesia, New Guinea, Australia and New Zealand. Most species grow as epiphytes but a few terrestrials are known. Growth habit, flowering patterns and floral details vary so much in this unwieldy group that it is impossible to generalise. Included however, are many of the most floriferous and colourful orchids grown in contemporary collections and even many of the less flamboyant types have become popular.

Cultivation: Because of the complexity of this genus it is not possible to cover all aspects of their cultivation. Generally however these orchids adapt well to cultivation and present few difficulties to growers. Some species can be grown well in pots but others prefer slabs of treefern, weathered hardwood or cork. Potting materials include particles of softwood bark, charcoal and fern fibre. Humidity should be high and a regular flow of fresh air is essential. Many species require bright light for good flowering. A number of species from mountainous regions are quite cold tolerant and grow well in temperate regions with only minimum protection. Species from lowland tropical areas will however require extra heat during winter.

Dendrobium amethystoglossum H. G. Reichb.

(with an amethyst labellum)

Although plants of this species have rather coarse stems to nearly one metre tall, this habit is more than offset by the beautiful drooping racemes of flowers which extend to about 12 cm long. Each raceme is densely flowered and bottlebrush-like with the prominent purple labellum of each flower contrasting prominently with the white sepals and petals. Of natural occurrence in the Philippines this species is mainly grown by collectors.

Temp	Shade	In/on	Humidity	Air Flow
Int-Hot	10-30%	Pot	30-50%	5

Dendrobium amethystoglossum

Dendrobium anosmum Lindley

(odourless)

A widely distributed species which extends from Malaysia to the Philippines, Indonesia and New Guinea. With its strongly pendulous pseudobulbs plants should be either attached directly to trees or slabs or grown in baskets. This species flowers best after a distinct dormant period when the new growths have just matured. Well grown plants develop pseudobulbs in excess of one metre long so plants need plenty of room to develop. This species is quite variable in flower size and colouration with superior variants mainly originating in the Philippines.

Temp	Shade	In/on	Humidity	Air Flow
Hot	10-30%	Slab or Basket	50-70%	5

Dendrobium arachnites H. G. Reichb.

(spider-like)

A remarkable orchid which is native to Burma, growing as an epiphyte near the border with Thailand. Although first collected in 1873 by William Boxall, this species is still a comparative novelty in collections. An excellent pot subject with its neat, tufted habit, the colourful flowers are borne in pairs or small groups in succession on a wiry raceme about 20 cm long. The brilliant red flowers have a large labellum prominently veined with purple. Plants are very cold sensitive and require high humidity with abundant air movement.

Temp	Shade	In/on	Humidity	Air Flow
Hot	30-50%	Pot	50-70%	5

Dendrobium anosmum

Dendrobium arachnites

Dendrobium capra J. J. Smith

(like a goat)

A little-known species from the Indonesian island of Java and often also reported as being native to Malaysia. Its interesting, greenish-brown, glossy flowers, about 3 cm across, are borne on slender racemes which arise among the upper leaves. The floral segments are of heavy texture and the flowers are long-lasting.

Temp	Shade	In/on	Humidity	Air Flow
Hot	10-30%	Pot	50-70%	5

Dendrobium capra

Dendrobium chrysotoxum Lindley

(a golden arch – in reference to the sprays of flowers)

A commonly grown species which is native to southern China, northern India, Burma and Thailand. Plants from lowland areas have lanky pseudobulbs whereas those from the mountains (up to 1000 m altitude) are much more stocky and better suited to cultivation. The highly attractive, long-lasting, fragrant flowers are borne on graceful, hanging or arching racemes to about 30 cm long. Well grown plants can carry a number of racemes, producing a very decorative floral display.

Temp	Shade	In/on	Humidity	Air Flow
Int-Hot	10-30%	Pot, Basket or Slab	30-50%	5

Dendrobium chrysotoxum

Dendrobium densiflorum (Wallich. ex Lindley) Kunze

(densely-flowered)

This species was much more popular with Australian growers earlier this century than it is today, although it is still a fairly commonly grown orchid. Plants of this species and *D. thrysiflorum* are similar and easily confused but when flowering are readily distinguished. *D. densiflorum* has yellow to orange flowers with a more prominently fringed labellum. This species grows on mossy tree trunks and branches in humid forests between 800 m and 1500 m altitude. It is native to northern India, Burma, Thailand and Laos.

Temp	Shade	In/on	Humidity	Air Flow
Int	10-30%	Pot	30-50%	5

Dendrobium fimbriatum Hook.

(fringed, in reference to the labellum margins)

Widely distributed in northern India, Burma, Thailand, Kampuchea, Laos and Vietnam, this species is popular for its buttercup-yellow flowers with a prominently fringed labellum. Plants are extremely easy to grow and flower regularly each year. A variant with one prominent dark blotch on the labellum is possibly more commonly grown than others. It is usually known by growers as var. *oculatum*.

Temp	Shade	In/on	Humidity	Air Flow
Cool-Int	10-30%	Pot or Slab	50-70%	4

Dendrobium densiflorum

Dendrobium fimbriatum

Dendrobium gibsonii

Dendrobium harveyanum

Dendrobium gibsonii Paxton

(after John Gibson, original collector)

Although this species has superficially similar flowers to the well-known *D. fimbriatum*, they are generally smaller, less fringed and readily recognisable by the two spots on the labellum (one in *D. fimbriatum*). Plants of both are similar but those of *D. gibsonii* have shorter, more slender stems. Although less popular with orchid growers than is *D. fimbriatum*, *D. gibsonii* is nevertheless a rewarding and easy subject to grow. It is fairly widely distributed in southern China, northern India and Burma.

Temp	Shade	In/on	Humidity	Air Flow
Int	30-50%	Pot	50-70%	5

Dendrobium harveyanum H. G. Reichb.

(after E. Harvey, English grower who first flowered the species)

The most striking feature of this orchid is the complex, branched fringing present on the margins of the petals and to a lesser extent the labellum. This imparts a fullness to the flower and also gives the false impression of being very hairy. Borne in small groups the bright yellow flowers have a pleasant honey fragrance most noticeable on warm days. The species is native to Burma and Thailand.

Temp	Shade	In/on	Humidity	Air Flow
Int-Hot	30-50%	Pot	50-70%	5

Dendrobium heterocarpum Wallich ex Lindley

(with fruit of more than one kind)

A vigorous orchid which produces aerial growths freely, making propagation a simple process. Plants grow readily and flower profusely, producing an attractive, long-lasting display. The widely opening flowers vary in colour from cream to greenish with the best clones being sulphur yellow with a prominent fuzzy labellum. Plants often tend to become untidy and are best grown in a basket or on a slab. This species is very widespread being common in India, much of South-east Asia, Indonesia and the Philippines. It is frequently grown under the synonym, *D. aureum*.

Temp	Shade	In/on	Humidity	Air Flow
Int	10-30%	Slab or Basket	50-70%	4

Dendrobium heterocarpum

Dendrobium infundibulum

Dendrobium infundibulum Lindley

(like a funnel, in reference to the shape of the labellum)

This orchid is popular in cultivation because it is very easy to grow and plants are more often in flower than not. A native of Burma and Thailand, it grows in the mountains on rocks and deciduous trees between 800 m and 1600 m elevation. The colour in the throat of the labellum varies from yellow to brick red. There is a popular notion among growers that plants with the latter colour all are of the variety *jamesianum* but this is incorrect. Those which are truly this variety have stouter, more rigid pseudobulbs and prominent small teeth on the inner surface of the lateral lobes of the labellum.

Temp	Shade	In/on	Humidity	Air Flow
Cool-Int	10-30%	Pot	50-70%	5

Dendrobium leonis (Lindley) H. G. Reichb.

(tawny yellow)

A common orchid found in lowland forests of Thailand, Malaysia, Borneo and Indonesia. Plants have unusual leafy stems and small, pale green or yellow flowers which have a pleasant vanilla-like perfume most noticeable on warm days. This species, which is mainly grown by ardent collectors, can be grown on slabs or in a small pot of coarse mixture.

Temp	Shade	In/on	Humidity	Air Flow
Int-Hot	10-30%	Pot or Slab	50-70%	5

Dendrobium leonis

Dendrobium loddigesii Rolfe

(after Conrad Loddiges & sons, 19th century orchid nurserymen of Hackney, England)

One of the most attractive small orchids in cultivation, this species is valued for its neat growth of short, prostrate or pendulous stems and the surprisingly large, colourful flowers. The leaves are shed quickly as the pseudobulbs fatten and mature, rendering the flowers even more conspicuous. The fragrant flowers are long lasting and as they age, the segments become papery and lighter in colour. This orchid is easily grown and specimen plants can provide an impressive floral display. Although for many years this species was believed to be native to India it is in fact of natural occurrence on the Chinese island of Hainan.

Dendrobium loddigesii

Temp	Shade	In/on	Humidity	Air Flow
Int	30-50%	Pot, Basket or Slab	50-70%	5

Dendrobium moschatum (Willd.) Sw.

(musk-scented)

This species makes a magnificent garden plant for tropical and subtropical regions. When attached to trees in suitable situations it forms a vigorous clump and the strongly striped, pendulous stems can reach more than 2 m in length. Floral displays are impressive but the musk-scented flowers last only about a week. In glasshouses of temperate regions this species makes an admirable basket plant. It is native to northern India, Burma, Thailand and Laos.

Dendrobium moschatum

Temp	Shade	In/on	Humidity	Air Flow
Int-Hot	10-30%	Basket or Slab	30-50%	5

Dendrobium nobile

Dendrobium nobile Lindley

(noble, stately)

This species is easily one of the most familiar and popular orchids in cultivation. A native of mountainous regions, it is of natural occurrence in southern China, Taiwan, northern India, Thailand, Laos and Vietnam. It is popular for its ease of culture, massed displays of long-lasting, large, colourful flowers and the range of variations exhibited in flower colour and form. The species first flowered in England in 1837 and it quickly became much sought after by collectors. It is one of the best orchids for specimen culture and there are records in English journals of large plants bearing more than one thousand flowers open at once. Such flowering specimens must have been a breathtaking sight indeed. Numerous variants of this species have been selected and many have been named as cultivars. Best growth and flowering of this orchid is achieved in temperate and subtropical regions. When conditions are unsuitable, plants flower poorly and produce an excess of aerial growths at the expense of flowers.

Temp	Shade	In/on	Humidity	Air Flow
Cool	10-30%	Pot, Basket or Slab	30-50%	5

Dendrobium pachyglossum Par. & H. G. Reichb.

(with a thick labellum)

Of natural occurrence in Malaysia, Laos, Vietnam, Burma and Thailand, this species grows in erect tufts on mossy rocks or on trees where the stems are usually pendant and may reach one metre long. Plants grow easily in warm, humid conditions but with its relatively inconspicuous flowers, this species is mainly of interest to the collector.

Temp	Shade	In/on	Humidity	Air Flow
Hot	30-50%	Pot, Basket or Slab	50-70%	5

Dendrobium pachyglossum

Dendrobium senile

Dendrobium senile Par. & H. G. Reichb.

(as if old)

A delightful orchid prized for its novelty value with its pseudobulbs and leaves being covered densely with long soft hairs. Plants grow quickly over summer and shed most of their leaves in winter after the pseudobulbs mature and prior to flowering. A native of altitudes at about 1000 m in the mountainous regions of northern Thailand, Burma and Laos, this orchid adapts well to cultivation. Plants will grow well mounted on a slab but seem to do better hanging high up in a small pot of fibrous mix.

Temp	Shade	In/on	Humidity	Air Flow
Int-Hot	10-30%	Pot or Slab	50-70%	5

Dendrobium thrysiflorum H. G. Reichb. ex Andre

(with a bunch-like inflorescence)

Plants of this species and *D. densiflorum* are difficult to distinguish from each other but at flowering are easily separated. The dense pendulous racemes of *D. thrysiflorum* are eye-catching and a substantial plant in flower is a beautiful sight to behold. With its white flowers and contrasting orange labellum, this species is difficult to confuse with any other. It is native to Burma, Thailand and Laos and grows at elevations up to 1200 m.

Temp	Shade	In/on	Humidity	Air Flow
Int	10-30%	Pot	30-50%	5

Dendrobium thrysiflorum

DISA Bergius

(Disa, a legendary Swedish figure)

A genus of about 120 species which are native to temperate and tropical parts of Africa, Madagascar and the Mascarene Islands. The majority of species grow as terrestrials with a few lithophytes preferring rock outcrops. These orchids have a tuber-bearing root system and distinct growth and dormant cycles. Ranging from lowland areas with a hot climate to high elevations where conditions are continually moist and humid, these orchids grow in a wide range of environments and many species have specialised cultural requirements.

Cultivation: As a group these orchids are considered difficult to grow and apart from a few adaptable species they are mainly the province of growers who are prepared to specialise. Well-drained moisture holding materials such as peat, perlite, sphagnum moss and vermiculite are commonly used for potting and many growers top-up the pots regularly with fresh live moss. When in active growth plants must be kept continually moist but as they begin to die down watering should be reduced and eventually witheld and the dormant tubers allowed to rest until new growth begins.

Disa uniflora Berg.

(single-flowered)

Generally regarded as one of the finest orchids in the world, until recently this species was only found in the collections of a fortunate few. It was also believed to be very difficult to grow but with strict adherence to basic techniques, modern growers achieve tremendous success. A native to the mountains of the south-western Cape in South Africa, this species grows beside streams and waterfalls in areas which are continually moist. Plants may even be submerged by running water during winter. The water is always cold and with an acidic pH. Air circulation is continual. Prized for its beautiful flowers, dedicated growers succeed admirably with these orchids. In nature plants rarely produce more than two flowers per spike, but cultivated plants often produce five or six flowers with up to twelve being recorded. Commonly the flowers are scarlet to bright red but pink, orange or yellow variants are known. It is interesting to note that *Disa uniflora* is easy to raise from seed (although successful pollination can be tricky to achieve). Seeds scattered on sterilised peat or sphagnum moss will begin to germinate in four to six weeks.

Temp	Shade	In/on	Humidity	Air Flow
Cool	30-50%	Pot	50-70%	5

Disa uniflora

DORITIS Lindley

(from the Greek, *dorys*, a spear; the labellum is spear-shaped)

This genus consists of a single species which is widespread in Asia. It is closely related to *Phalaenopsis* but with a very distinctive growth habit and different floral features.
Cultivation: Usually this species is grown in a pot in a mixture of fine particles of bark, charcoal and fern fibre. Plants need plenty of warmth, abundant air movement and bright light to flower and year-round, regular watering. They also respond to fortnightly fertilising during summer.

Doritis pulcherrima Lindley

(very beautiful)

An extremely variable species with flowers ranging in size from 1-4 cm across and in colour from white through light pink to deep mauve. Because it is vigorous and easy to grow, this species is popular with orchidists and a number of superior clones have been selected for cultivation. Clumps increase in size fairly rapidly and well-grown plants are in flower over much of the year. The stiffly erect flower stems can reach one metre tall and under good conditions may continue producing flowers for as long as they elongate. Best flowering is achieved in bright light. A native of the hot, humid, lowland jungles of South-east Asia, Malaysia and Indonesia, this species is very popular in Asia where it is almost grown as a garden plant. The close relationship between *Doritis* and *Phalaenopsis* is illustrated by the fact that *D. pulcherrima* has been described on at least four occasions as a *Phalaenopsis*, the best known being *P. esmeralda*.

Temp	Shade	In/on	Humidity	Air Flow
Hot	10-30%	Pot	70-90%	5

Doritis pulcherrima

Encyclia ambigua

ENCYCLIA Hook.

(from the Greek *encyclein*, to encircle; a reference to the labellum which encloses the column)

The taxonomy of this genus has been intimately tied in with that of *Epidendrum* and only recently have the differences between each genus been clarified. Most species of *Encyclia* have prominently swollen, pear-shaped pseudobulbs whereas reed-like stems predominate in *Epidendrum*. This distinction is not hard and fast, however, because a few species of *Epidendrum* have pseudobulbs and a couple of *Encyclia* species are reed-stemmed. In the flowers of *Epidendrum* the column and base of the labellum is completely united whereas in *Encyclia* these organs are free or partially united. A deep slit, left in the top of the column after removal of the pollinarium, is prominent in *Epidendrum* but absent in *Encyclia*.
Encyclia numbers about 150 species which are predominantly found in Mexico and the West Indies. They grow on rocks or trees in a wide range of positions from shade to nearly full sun.
Cultivation: Many species of *Encyclia* are well established in cultivation; a number being very easy to grow and with attractive flowers are firm favourites with growers. Most species are grown in pots and plants should be underpotted rather than overpotted. Suitable potting materials include softwood bark, charcoal and fern fibre. Some species grow well on slabs. All of these orchids are evergreen but most have a distinct period of dormancy when watering should be much reduced. Actively growing plants need daily watering and fertilising every two or three weeks. Although tolerant of low light most species grow and flower best in fairly bright light. Air movement should be free and unimpeded.

Encyclia ambigua Schltr.

(doubtful, uncertain)

This species is included by some authorities as a synonym of *E. alata* but it has larger flowers (about 3.5 cm across) and a broader labellum which is spotted and veined with red. A native of Mexico and Guatemala, this species grows on trees in sparse forests at low altitudes and sometimes as a terrestrial in accumulations of humus. Cultivated plants flower freely and the fairly long-lasting flowers are delicately fragrant.

Temp	Shade	In/on	Humidity	Air Flow
Int	10-30%	Pot or Slab	30-50%	5

Encyclia atropurpurea Schltr.

(dark purple)

A widespread species which occurs naturally in Mexico, Cuba, Central America and northern countries of South America. It grows as an epiphyte at low to moderate elevations in dense, low scrub. With its large, colourful flowers and adaptability to cultivation, this highly ornamental species deserves to become much more widely grown. The long-lasting, fragrant flowers are variable in colour. One attractive variant has a brilliant magenta labellum. Some authorities treat *E. atropurpurea* as a synonym of *E. cordigera*.

Temp	Shade	In/on	Humidity	Air Flow
Int	10-30%	Pot	30-50%	5

Encyclia cochleata

Encyclia atropurpurea

Encyclia cochleata (L.) Lemee

(like a shell)
Clamshell Orchid

This species has the honour of being the first tropical epiphytic orchid to flower at Kew Gardens, England, that event taking place in 1787. A native of Florida, Mexico, Central America, South America and the West Indies, this species is a firm favourite in cultivation. In nature plants grow in a wide variety of habitats, in both deciduous and evergreen forests, from sea level to about 2000 m elevation. When flowering, the broadly striped (almost black), broadly heart-shaped labellum is thrust upwards while the narrow, yellowish-green sepals and petals hang downwards, the whole structure projecting the image of some exotic sea creature.

Temp	Shade	In/on	Humidity	Air Flow
Int	10-30%	Pot or Slab	50-70%	5

Encyclia cordigera (Kunth.) Dressler

(with heart-shaped organs – in reference to the labellum)

A beautiful species often mistakenly grown as *Epidendrum atropurpureum*. Although the sepals and petals are dull brown the lovely pink labellum provides a colourful contrast. The species is native to Mexico, Panama, Venezuela, Peru and the West Indies, growing from near sea level to about 900 m elevation in the mountains.

Temp	Shade	In/on	Humidity	Air Flow
Int-Hot	10-30%	Pot	50-70%	5

Encyclia cordigera

Encyclia dichroma (Lindley) Schltr.

(of two colours)

In Brazil, where this fine species is native, *Encyclia dichroma* grows on the branches of low straggly shrubs and its roots often extend into the sandy soil of the surrounding area. The prevailing climate is strongly seasonal and the plants must be able to withstand dry periods of two or three months or even longer. This climate seasonality is reflected in the growth habit of this orchid which has a strong dormant period during which plants must be kept on the dry side.

Temp	Shade	In/on	Humidity	Air Flow
Int-Hot	10-30%	Pot or Slab	30-50%	5

Encyclia mariae (Ames) Hoehne.

(derivation unknown)

A lovely miniature orchid with greyish pseudobulbs less than 5 cm long, longer leaves (to 18 cm long) and disproportionately large flowers which are about 7 cm long. The flower is unusual because it is dominated by the huge, flared, frilly white labellum which contrasts markedly with the narrow green sepals and petals. A native of Mexico, this species grows in cool, humid forests at about 1000 m altitude.

Temp	Shade	In/on	Humidity	Air Flow
Cool-Int	30-50%	Pot	50-70%	5

Encyclia dichroma

Encyclia mariae

Encyclia pentotis (H. G. Reichb.) Dressler

(in fives)

Although related to *E. cochleata*, this species has a much smaller labellum and presents its flowers in a vastly different manner. Of natural occurrence in Mexico, Guatemala, El Salvador, Honduras and Brazil, this species is very common in the highlands extending to about 1700 m altitude. It grows as an epiphyte on trees in humid forests. Some authorities include this species as a synonym of *Encyclia baculus*.

Temp	Shade	In/on	Humidity	Air Flow
Int	10-30%	Pot	50-70%	5

Encyclia pentotis

Encyclia polybulbon Sw.

(with many bulbs)

A most attractive orchid with a compact but spreading growth habit and small but colourful flowers produced at intervals throughout the year. Of natural occurrence in mountainous regions of Mexico, Guatemala, El Salvador, Honduras, Cuba and Jamaica, this orchid grows as an epiphyte at altitudes of about 1200 m, usually in situations of bright light and with plenty of air movement. It makes an attractive basket specimen and can also be grown happily on a slab. It is still commonly grown and sold as *Epidendrum polybulbon* and is sometimes classified in the genus *Dinema*.

Temp	Shade	In/on	Humidity	Air Flow
Int	10-30%	Basket or Slab	50-70%	5

Encyclia polybulbon

Encyclia prismatocarpa (H. G. Reichb.) Dressler
(with three sharp angles on the fruit like a prism)

Highland areas of the Central American countries Costa Rica and Panama are the home of this attractive orchid. Often locally common on trees and shrubs, it has been recorded growing in forests up to 3300 m altitude. Plants usually have a yellowish appearance even when protected from strong light and the long-lasting, colourful flowers are borne on arching racemes and have an attractive perfume.

Temp	Shade	In/on	Humidity	Air Flow
Cool-Int	10-30%	Pot	50-70%	5

Encyclia radiata (Lindley) Dressler
(spreading like rays)

A species of low to moderate elevations (to 1200 m) in Mexico, Costa Rica and Venezuela where it grows on trees and rocks in open to dense forests. Plants have clusters of strongly ribbed, yellowish pseudobulbs and erect, dense clusters of strongly fragrant, heavily textured flowers in which the erect, shell-shaped labellum has prominent purple stripes. The flowers are long lasting and plants grow readily in cultivation.

Temp	Shade	In/on	Humidity	Air Flow
Int-Hot	10-30%	Pot	50-70%	5

Encyclia radiata

Encyclia prismatocarpa

*Encyclia
tampensis*

Encyclia tampensis (Lindley) Small

(from Tampa Bay, near Florida, USA)
Florida Butterfly Orchid

A very common orchid which last century was reported as growing in huge colonies in southern Florida and is still reasonably common today. Originally believed to be an American endemic, it is now also known from the Bahamas and a variant (var. *amesianum* Correll) occurs in the mountains of Cuba. Plants grow in huge, dense mats and the long-lasting flowers are pleasantly fragrant. This species is very easily grown and in subtropical climates makes an attractive outdoor plant on garden trees.

Temp	Shade	In/on	Humidity	Air Flow
Cool-Int	10-30%	Pot, Basket or Slab	50-70%	4

EPIDENDRUM L.

(from the Greek, *epi*, upon; *dendros*, tree; in reference to the epiphytic habit)

A genus of about 700 species found in southern states of the United States of America, Mexico, Central America, South America and the West Indies. One of the largest genera of orchids, it contains some diverse elements, however, the recent re-recognition of *Encyclia*, has removed some of that diversity. Most species of *Epidendrum* have slender, reed-like stems that are not swollen into pseudobulbs and alternate leaves arranged on opposite sides of these stems. A few species however have typical, swollen pseudobulbs. The base of the column and labellum of the flowers being united and a prominent slit in the top of the column after removal of the pollinarium are further useful features for identifying members of this genus. A considerable number of *Epidendrum* species are to be found in contemporary collections and they are popular for their ease of culture and interesting, colourful flowers. In the tropics, colourful species and hybrids of some reed-stemmed types are commonly grown as garden plants.

Cultivation: The reed-stemmed types so commonly seen in tropical gardens can be grown in loose sandy soil but are more usually situated among rocks, coral or coconut husks. Most other species are grown in pots and the bulk of them require the warmth and protection of a glasshouse. Potting mixtures can be made up from softwood bark, charcoal, perlite and fern fibre. A few species grow well on slabs of cork or treefern. Many of these orchids seem to grow almost continually throughout the year, some at best having a short rest period. Composts should be kept moist throughout the year with watering being reduced over winter. Light levels should be fairly bright for flowering but many species can tolerate low light. Adequate air movement is important.

Epidendrum ciliare L.

(fringed as with eyelashes)

This species has been cultivated in Europe for nearly two hundred years, being introduced to England from the West Indies in 1790. It is also native to Mexico and a number of countries in Central America and South America. Plants grow on rocks and trees from the lowlands to about 2000 m elevation. One of a group in the genus which has stems distinctly swollen into pseudobulbs, plants of this popular species may be easily mistaken for *Cattleya labiata* when not in flower.

Temp	Shade	In/on	Humidity	Air Flow
Int	10-30%	Pot or Slab	50-70%	5

Epidendrum ibaguense

Epidendrum ibaguense Kunth

(from the area of Ibague, Colombia)
Crucifix Orchid

The reed-like stems of this common, widespread species have been measured up to 10 m in length and may be supported by surrounding vegetation in a similar manner to some climbing plants. It is native to Mexico, Central America and South America. With its ease of culture and dense, well-displayed clusters of colourful flowers, this species is an ideal orchid for beginners and is also a familiar garden plant. Flower colour is usually scarlet or red but orange, yellow or bi-coloured variants are known. The distinct cruciform labellum gives rise to the common name of Crucifix Orchid. Very common in tropical and subtropical areas, this species is hardy enough to be grown in a warm position as far south as Melbourne.

Temp	Shade	In/on	Humidity	Air Flow
Cool-Hot	10-30%	Pot	30-50%	5

Epidendrum ciliare

Eria xanthocheila

ERIA Lindley

(from the Greek, *erion*, wool; a reference to the hairy flowers of this genus)

Estimates suggest that this genus numbers between 350-500 species, which makes it a highly significant group of orchids. Most grow as epiphytes in humid forests with some being lithophytes and very few growing in the ground. Their hairy flowers are generally short-lived but they are commonly produced in profusion and are often colourful. Some species may flower more than once a year. Very few species have become entrenched in cultivation.

Cultivation: Species of *Eria* are generally easy to grow providing their basic requirements are met. With their very fine roots, plants are sensitive to poor drainage, lack of aeration and clogging of the potting mix. They also dislike being overpotted. Suitable potting materials included fern fibre, bark, charcoal and chopped sphagnum moss. Many species will also grow well on slabs. Watering should be copious while in active growth and reduced when the plants are dormant. Gentle air movement and high humidity are very beneficial. Those species originating from the lowlands are very sensitive to cold and can collapse dramatically after such exposure.

Eria xanthocheila Ridley

(with a yellow lip)

Distributed in Malaysia, Indonesia and Borneo, this species grows on trees along streams in warm, humid lowland forests. Plants have elegant, slender pseudobulbs crowned with spreading leaves and at flowering produce dense inflorescences of crowded colourful flowers. These have a conspicuous yellow labellum which gives rise to the specific epithet. A rewarding, fairly easily grown species.

Temp	Shade	In/on	Humidity	Air Flow
Hot	30-50%	Small Pot or Slab	50-70%	5

EULOPHIA R. Br.

(from the Greek *eu*, well; *lophos*, a crest; these orchids have a crest-like callus on the labellum)

A genus of about 250 species of orchids which are widely distributed in the tropical regions of the world with a significant development in Africa. The majority of species are terrestrials but rare lithophytes and epiphytes are known. Many species have limited ornamental qualities but some exceptional species have large or colourful flowers.

Cultivation: Species of *Eulophia* are grown in pots of a soil-based mixed (good loam, coarse sand, charcoal, leaf mould) or a finely-particled fibrous mix (fern fibre, pine bark, charcoal). They have a strong dormant period when leafless and should be kept dry until the onset of new growth. Plants are watered copiously when in active growth. The use of old animal manure is beneficial.

Eulophia keithii Ridley

(derivation unknown)

Distributed in Thailand and Malaysia, this species is reported to grow on limestone rocks in areas with a seasonal climate. Plants are leafless during a long dormant period in the dry season, producing new growth and flowering with the onset of the wet. The tall flower spikes arise from the base of a recently mature pseudobulb together with the new shoots. This species adapts very well to cultivation and plants flower over a long period. Limestone chips are beneficial in the potting mix.

Temp	Shade	In/on	Humidity	Air Flow
Int-Hot	10-30%	Pot	30-50%	4

Eulophia keithii

Eulophia streptopetala Lindley

(with twisted petals)

An African orchid which is widely distributed over the continent from Ethiopia to South Africa. It grows in a wide range of soil types usually in patches of vegetation where the trees afford some shade. This orchid has coarse, thick roots, prominent dark green pseudobulbs and large, strongly-ribbed, thin-textured leaves. The flower spikes arise from the base of a recently mature pseudobulb, usually coinciding with the appearance of new growth. The highly colourful flowers are long lived and last well when cut. this species is sometimes grown under the synonym, *E. krebsii*.

Temp	Shade	In/on	Humidity	Air Flow
Int-Hot	10-30%	Pot	30-50%	5

Eulophia streptopetala

GALEANDRA Lindley

(from the Latin, *galea*, a helmet; *andros* male; in reference to the helmet-shaped anther cap)

A genus of approximately 25 species all of which are found in regions between South Florida and Brazil in South America. They grow mainly as epiphytes in open forest but some species have been recorded as growing in accumulations of litter on the ground. Plants have a distinct growing stage after which the pseudobulbs harden and flowers are produced. After leaf fall the plants are dormant for a period of four to eight weeks prior to the onset of new growth.

Cultivation: Species of this genus are vigorous growers but are sensitive to cold. In general they require warm humid conditions with free air movement and bright light. When dormant, plants should be watered sparingly. New growths are susceptible to rotting if water lodges in the upper expanding leaves. Mostly these orchids are grown in pots using bark, charcoal and fern fibre as a growing mix.

Galeandra batemanii Lindley

(after James Bateman, 19th century English orchidist)

Although this orchid is often grown as *G. baueri* that species is quite distinct being recognised by its yellow flowers. *G. batemanii* is native to Mexico and Central America where it grows on trees in open forest. With its large colourful, long-lasting flowers this species is fairly popular in cultivation. Plants grow readily in a pot and flower quite freely.

Temp	Shade	In/on	Humidity	Air Flow
Hot	10-30%	Pot	50-70%	5

Galeandra batemanii

Galeandra devoniana Lindley

(after the Duke of Devonshire, English orchidist)

In nature this is a vigorous orchid which can develop into clumps 3 m or 4 m across and with pseudobulbs up to 2 m long. Of more modest proportions, cultivated plants nevertheless grow strongly and bear their large, handsome flowers freely. These have a very prominent white labellum the upper surface of which has a delicate filigree of purple markings.

Temp	Shade	In/on	Humidity	Air Flow
Hot	10-30%	Pot	50-70%	5

Galeandra devoniana

GASTROCHILUS D. Don

(from the Greek, *gaster*, a belly; *cheilos* lip; in reference to the deeply pouched labella of these orchids)

This genus is comprised of about 20 species of small epiphytes which are distributed in India, South-east Asia and Malaysia. Small monopodial orchids, they have a similar growth habit to species of *Sarcochilus* but with very distinctive flowers. The labellum is deeply saccate with fringed, lacerated or hairy margins. The flowers themselves are often colourful, usually fragrant and long lasting. In most species they are crowded on short racemes and this enhances the floral effect.

Cultivation: These small orchids are best mounted on slabs of treefern, English Oak or cork. Root growth is usually vigorous when the plants are healthy. Growing requirements are high humidity, diffuse light and continuous gentle air movement. Watering should be regular throughout the year with a reduction in winter as growth slows down. Mealy Bugs and White Fly can cause severe problems with these orchids.

Gastrochilus bellinus

Gastrochilus bellinus (H. G. Reichb.) Kuntze

(beautiful, pretty)

A lovely, small orchid from the rainforests of Thailand and Burma where it grows in the lowlands and at moderate altitudes. Plants have a neat habit typical of many small species of Sarcanthinae but with very distinctive flowers. These fleshy, widely opening flowers, each about 3 cm across, radiate stiffly from short racemes to present a crowded appearance. They are long-lived, delightfully fragrant and with a prominent, white fringed labellum.

Temp	Shade	In/on	Humidity	Air Flow
Int-Hot	30-50%	Slab	50-70%	5

GOMESA R. Br.

(after Bernardinus Antonius Gomes, author of a book on medicinal plants of Brazil)

Species in this genus were originally described in *Rodriquezia*, but in that genus the lateral sepals are united and folded together to form a platform under the labellum or recurved to resemble a spur. In *Gomesa* the sepals may be loosely united but are never folded together as a unit. *Gomesa* consists of about 10 species and all appear to be endemic to Brazil where they grow on trees or rocks in equatorial forests. Plants of this genus are generally uncommon in cultivation. The flowers, although small, are long lived and attractively displayed on arching racemes.

Cultivation: Being rather cold sensitive, these orchids require warm to hot, humid conditions with regular gentle air movement. They are tolerant of fairly low light intensities and need regular watering throughout the year as they have only a short rest period after growths have matured. They can be successfully grown in pots of fern fibre or bark.

Gomesa crispa (Lindley) Kl. & H. G. Reichb.

(curled closely or crested)

This species is similar in general habit to *G. recurva* but the lateral sepals spread widely apart and the segments have more prominently crisped margins. Native to the coastal mountains of southern Brazil, this orchid has a neat growth habit with flattened, sharply-edged, pale green pseudobulbs and a loose fan of relatively broad leaves. The green flowers are regularly spaced on arching racemes which may reach 20 cm in length. Each pseudobulb may produce several racemes and large flowering plants are showy.

Temp	Shade	In/on	Humidity	Air Flow
Hot	30-50%	Pot or Slab	50-70%	5

Gomesa crispa

GONGORA Ruiz. & Pav.

(after Don Antonio Caballeroy Gongora, one time Viceroy of New Granada)

A genus of about 20 species of epiphytic orchids with rather small flowers of a most complex structure. The flowers, which hang upside down on fairly long racemes, have very short petals which project forwards beside the column (which is like a long, waxy slide) and a horizontal labellum of most unusual shape. The plants have clustered, strongly ridged pseudobulbs and a pair of broad, fairly thin-textured leaves. They are native to various countries in Central America and South America.

Cultivation: With their pendulous racemes these orchids are commonly grown in baskets, although they will also do well on slabs and in hanging pots. Fibrous mixes of fern fibre and chopped sphagnum moss are successfully used by many growers, but others have success with softwood bark, charcoal, chopped moss and fibre. Drainage must be excellent and gentle, unimpeded air movement is essential. As some species originate from fairly high altitudes they can be fairly tolerant of cold.

Gongora armeniaca (Lindley) H. G. Reichb.

(apricot-coloured or perhaps scented)

A native of Nicaragua, Cost Rica and Panama, this species grows as an epiphyte at elevations up to 1200 m. With its yellow, orange or salmon-coloured flowers (among the largest in the genus) on long pendent racemes, this species makes an interesting basket subject. Well grown plants produce impressive floral displays. The flowers also release an attractive perfume that is somewhat reminiscent of apricots.

Temp	Shade	In/on	Humidity	Air Flow
Int-Hot	10-30%	Basket or Slab	50-70%	5

Gongora armeniaca

Gongora galeata (Lindley) H. G. Reichb.

(as if wearing a helmet)

A most impressive species when in flower, this orchid produces pendent racemes of rather large pale yellow or brownish flowers. Each flower is borne at the end of a prominent incurved pedicel and ovary, these structures almost imparting the impression of the links of a chain. The upwards-facing flowers are long-lasting, fragrant and are displayed to their best advantage from a hanging container.

Temp	Shade	In/on	Humidity	Air Flow
Int-Hot	10-30%	Basket or Slab	30-70%	5

Gongora galeata

GRAMMATOPHYLLUM Blume

(from the Greek, *gramma*, letter; *phyllon*, a leaf; alluding to the conspicuous markings on the perianth parts)

Orchids of this genus are large, in fact some species are best described as massive because mature clumps are of dimensions that would fill the tray of a dump truck and weigh in excess of two tonnes. Pseudobulbs of *G. speciosum* have been measured at about 8 m long and the racemes can be of a similar length. The large, long-lasting flowers are invariably heavily blotched and marked. An unusual feature is that the basal flowers of a raceme are usually abnormal and may be unisexual. Plants also have a distinctive aerial root system which acts as a litter-collecting device.

Cultivation: Orchids of this genus are easy to grow but must be given plenty of room to reach their potential. They require warm to hot conditions, bright light and fairly high humidity with plenty of air movement as they are very sensitive to bacterial rots. Watering should be copious over summer and reduced in frequency as the pseudobulbs mature. Drainage should be perfect and fairly coarse potting materials are commonly used. These orchids are gross feeders and should be fertilised regularly during their growing cycle.

Grammatophyllum scriptum (L.) Blume

(as though written on)

In their native state these orchids grow on trees and rocks in humid forests in conditions of bright light and where air movement is almost constant. Found naturally in the Philippines, Borneo, Celebes, Moluccas, Solomon Islands and New Guinea, they form large congested clumps. A mass of white roots forms an intricate erect network around the pseudobulbs, this acting as a litter-trapping device which gathers nutrients to the plant. This species is easily grown and very rewarding in cultivation.

Temp	Shade	In/on	Humidity	Air Flow
Hot-VHot	10-30%	Pot	30-50%	5

Grammatophyllum scriptum

GROBYA Lindley

(after Lord Grey of Groby, 19th century English patron of Horticulture)

A small genus of about 4 species all endemic to Brazil. Rare in cultivation, these orchids grow naturally as epiphytes in the warm, humid forests of the lowlands.

Cultivation: These orchids are best grown in a pot of fern fibre or a mixture of softwood bark, charcoal and sand. They require warm, humid conditions with an abundance of air movement. These orchids should never be allowed to dry out completely however plants have a short quiescent period prior to flowering when they should be kept on the dry side.

Grobya amberstiae Lindley

(after Lady Amherst)

This species can be recognised from others in the genus by its ovoid pseudobulbs, its broad petals which are blotched like a leopard and the prominent lobe on the apex of the labellum. The attractive flowers are fairly long-lived and have a pleasant perfume. A native of Brazil, this interesting species is rare in cultivation, mainly being an item for the very keen collector.

Temp	Shade	In/on	Humidity	Air Flow
Hot	30-50%	Pot	50-70%	5

Grobya amberstiae

HABENARIA Willd.

(from the Latin *habena*, rein or strap; in reference to the slender to filiform lateral lobes on the labellum of some species)

A very large genus of 600-700 species which are widely distributed in the tropical regions of the world. Most are terrestrial orchids with a deciduous habit, dying back to subterranean tuberoids to avoid extremes of dryness. A very few species are epiphytes. They occupy a tremendous range of habitats and soil types but often frequent soils which are inundated for some season during the year. Floral morphology is extremely variable and flower size ranges from minute to flamboyant and they are often arranged in dense racemes. Colours are usually dull (greens, cream and white) but a few species are highly colourful. Many species are highly fragrant. Because they have a long dormant period these orchids are mainly for the specialist and very few are grown in mixed collections.

Cultivation: During the dormant period, the tuberoids of these orchids must be kept dry. If watered at this stage, rotting can occur. When the new shoots appear above the soil, watering begins and the plants are kept regularly moist while in growth. As the plants die back watering is again reduced. Repotting takes place while the tuberoids are dormant. A terrestrial mix similar to that outlined in chapter 4 is suitable. Plants should be grown in bright light with warmth, humidity and air movement. Many species are very sensitive to cold.

Habenaria rhodocheila Hance

(with a red labellum)
Red Man Orchid, Little Dragon Orchid

A lovely orchid which grows in shady areas near streams and also in more exposed sites around rocks and even in crevices of litter on the rocks themselves. Plants have attractive, deep green leaves which are often somewhat crinkled and large colourful flowers which light up the gloom. These vary tremendously in colour including cream, pink, yellow, orange, red and scarlet. This species is one of the most popular members of the genus in cultivation and it grows well. It is native to China, Thailand, Laos, Vietnam, Hong Kong, Malaysia and the Philippines.

Temp	Shade	In/on	Humidity	Air Flow
Int-Hot	30-50%	Pot	50-70%	5

Habenaria rhodocheila

HUNTLEYA Batem. ex Lindley

(after Rev. Mr. J. T. Huntley, early 19th century English orchidist)

A genus of about 5 species of pseudobulbless epiphytes which are found naturally at moderate elevations in Central America and South America. Species of *Huntleya* have at various times been included in other genera such as *Zygopetalum* and *Batemania*. Plants have an interesting fan-like growth habit and highly decorative flowers. Only one species is grown to any extent and even that species is a collector's item.

Cultivation: These orchids have a reputation of being tricky to grow but success can be achieved if they are given cool, moist conditions with plenty of air movement. If however these conditions are not met the plants either linger or quickly die. Watering should be regular over most of the year but reduced as plants slow down over winter. Fern fibre seems to be the most suitable potting material.

Huntleya meleagris Lindley

(spotted like a Guinea Fowl)

The magnificent flowers of this orchid are amongst the most beautiful in the family. Spreading to about 12 cm across, they open widely, are pleasantly fragrant, long lasting and with a waxy, almost polished lustre. Plants will not flower unless they are happy with their growing conditions and a display like that produced on the plant illustrated is a sure sign of optimum conditions. Being a plant from mountainous regions, this species likes relatively cool conditions with an abundance of cool, moist air to counteract the heat of summer. Bright light is not essential for flowering.

Temp	Shade	In/on	Humidity	Air Flow
Int	30-50%	Pot	70-90%	5

Huntleya meleagris

IONOPSIS Kunth

(from the Greek *ion*, violet; *opsis*, appearance; the flowers bear a fanciful resemblance to violets)

A small genus of about 5 species distributed in Florida, Mexico, Central America and the Galapagos Islands. They are small epiphytes with attractive flowers borne on wiry, branched inflorescences. Only one species is commonly grown.

Cultivation: It is common for cultivated plants of this genus to decline steadily, especially if insufficient attention is given to air movement. Plants can be grown on slabs or in small pots and should be hung close to the roof of the glasshouse. Prevailing conditions should be warm and humid. In the tropics, plants are best grown on live trees, especially varieties of citrus.

Ionopsis utricularioides (Sw.) Lindley

(like the genus *Utricularia*)

The small, colourful flowers of this orchid are dominated by the large, lobed labellum and are carried in abundance along arching wiry panicles in a similar manner to many species of *Oncidium*. Usually an epiphyte on trees in tropical lowlands, this species is known to colonise cultivated citrus trees in some countries. It occurs naturally in Florida, Mexico, Bolivia, Paraguay, Brazil and the Galapagos Islands.

Temp	Shade	In/on	Humidity	Air Flow
Int-Hot	10-30%	Pot or Slab	50-70%	5

ISABELIA Barb. Rodr.

(after Isabel, early Brazilian Protectress of science)

A genus of a single species which is endemic to Brazil. Although this orchid is very attractive it is mainly found in the collections of ardent enthusiasts.

Cultivation: This orchid, which is relatively easy to grow, has a strong creeping rhizome which branches freely and forms a mat of compact growth. Plants are best grown on a slab or in a shallow tray and require warm, humid conditions with plenty of gentle air movement. Fern fibre or fine particles of bark are suitable for potting.

Isabelia virginalis R.Br.

(pure white)

A native of Brazil this miniature orchid forms dense spreading mats on mossy branches in humid rainforests. Crystalline-white, semi-tubular flowers, each only about 6 mm long, contrast with the mat of dark green foliage. The flowers are not long-lasting but a well-grown plant may flower intermittently throughout the year with a flush in winter-spring. A novelty orchid mainly of interest to the collector. Plants are grown readily in a shallow pot or saucer of relatively fine-particled epiphyte mix or on a slab of hardwood or treefern. They need shade, humidity combined with good air movement, regular watering throughout the year and warmth in winter.

Temp	Shade	In/on	Humidity	Air Flow
Int	50-70%	Pot or Slab	70-90%	5

Ionopsis utricularioides

Isabelia virginalis

ISOCHILUS R. Br.

(from the Greek *isos*, equal; *cheilos*, a lip; the labellum is equal in size to the sepals)

A small genus of only 2 species but each of these is variable in the extreme in such features as leaf shape and arrangement and flower size and colour. Being very rewarding subjects to grow they offer considerable scope for selection of superior forms and the display of colour variants.

Cultivation: Excellent orchids for the beginner, species of *Isochilus* are easy to grow and quickly build up into a potful They can be grown in pots and baskets or on slabs of treefern cork, English Oak or weathered hardwood. Suitable potting materials include fern fibre and bark. Continual gentle air movement and warm, humid conditions are recommended.

Isochilus linearis (Jacq.) R. Br.

(linear, in reference to the leaves)

A rewarding species which quickly develops into a tufted clump and regularly produces its colourful, starry flowers from the ends of the leafy stems. Being extremely variable in colour (white to mauve, purple, orange, yellow, red or pink) this species offers scope for selection and is a very collectable item. In nature this species grows in a variety of ways including as a terrestrial, lithophytic on rocks, on rotting logs and epiphytic on trees. It is native to Mexico, Central America, South America and the West Indies and is distributed from near sea level to about 4000 m altitude.

Temp	Shade	In/on	Humidity	Air Flow
Int-Hot	10-30%	Pot or Slab	50-70%	5

Isochilus linearis

KEFERSTEINIA H. G. Reichb.

(after Herr Keferstein, 19th century German orchid grower)

This unusual genus consists of about 20 species, all being confined to Central America and South America. Plants lack pseudobulbs and have their leaves in a spreading fan. The relatively large, colourful flowers are borne singly on slender racemes arising from the base of a growth. Essentially species of shady humid rainforests at moderate to high elevations, these attractive orchids are rarely encountered in cultivation.

Cultivation: Plants of these orchids have no capacity to withstand dryness and should be kept moist throughout the year. They also need shade, high humidity and an abundance of gently moving air. Mostly they are grown in pots of a fibrous mix using such materials as fern fibre, bark, charcoal and chopped sphagnum moss.

Kefersteinia tolimensis Schltr.

(from the area of Tolimen)

Highland rainforests of Venezuela, Colombia and Ecuador at about 2000 m elevation are the natural home of this delightful orchid. Humidity is constantly high throughout the year, temperatures moderate and the plants are subject to continual air movement interspersed with occasional dampings from mists and clouds. Plants grow perched on the branches of trees in shady positions. Although strictly a collector's item this species grows quite well in conditions similar to those which are successful for *Odontoglossum*.

Temp	Shade	In/on	Humidity	Air Flow
Int-Hot	50-70%	Pot	70-90%	5

Kefersteinia tolimensis

Laelia anceps

LAELIA Lindley

(after Laelia, one of the Vestal Virgins)

To many growers the differences between this genus and *Cattleya* seem largely superficial. Although the small, colourful types of *Laelia* are readily discernible, the flamboyant species such as *L. purpurata* seem to merge in with the other genus. The only stated distinction between the genera is to be found in the pollinia but even here there are exceptions; *Cattleya* has four pollinia arranged in a single series (but *C. dormaniana* has eight in two series of four) whereas *Laelia* has eight in two series of four; in some species of *Laelia* four of the pollinia are much smaller than the others.

Whatever the generic affinities, the genus *Laelia* includes some of the loveliest of all cultivated orchids. Distributed in Mexico, the West Indies, Central America and South America as far south as Argentina, the genus numbers about 80 species. An epiphytic or lithophytic growth habit predominates but the growth habit is variable.

Some species of *Laelia* have elongated pseudobulbs and large leaves as in *Cattleya*; others have short, clustered pseudobulbs. Floral variation is amazing with some species having solitary flowers on short racemes whereas others have clustered flowers on racemes which can be more than 2 m long. A very distinctive group of species has been given section status within the genus (section *Parviflorae*) and these are commonly known by growers as the 'rupicolous' laelias. These species are mostly rock dwellers and are characterised by short, crowded pseudobulbs, thick leathery leaves and small but highly colourful flowers. About 35 species of rupicolous laelias are known and most occur naturally on the mountains of Brazil. Heavy early morning dews enable these orchids to survive long dry winters on otherwise arid rocks in exposed sunny conditions.

Cultivation: Apart from members of the section *Parviflorae*, most species of *Laelia* take well to cultivation, many being adaptable enough for mixed collections, others growing well with cattleyas. Species originating in the mountains are cold hardy whereas those from the tropics need much warmer conditions. Free air movement, high humidity and good light for flowering are necessary for successful cultivation. Most species are grown in pots and suitable potting materials include softwood bark, fern fibre and charcoal. Hardy species, such as *L. anceps*, can be grown on slabs and also can be attached very well to garden trees in temperate and subtropical regions.

The rupicolous laelias have rather specialised cultural requirements arising no doubt from their unusual habitats. They do not adapt well to a general mixed collection and are best grown on their own. Plants are extremely sensitive to poor drainage and afflicted roots rot quickly starting a decline in the general health of the plant. Overwatering and excessively fine particles in the potting mix can induce the same result. Successful growers of these orchids use small pots (usually terracotta) and coarse potting materials (chunks of sandstone, softwood bark and charcoal). Air movement should be continual and bright light is necessary for flowering. Watering should be regular while plants are in active growth but is reduced after the pseudobulbs have matured until the onset of new shoots. Repotting should be carried out immediately the mix shows signs of deterioration.

Laelia anceps Lindley

(winged, in reference to the flattened inflorescence)

First introduced into England in 1835, this species has become a firm favourite with orchid growers throughout the world. It is prized for its ease of culture and free-flowering nature. The flowers open widely, are long-lasting and are typically rosy purple with a darker lip boldly marked with yellow. They are borne on a graceful spike which may reach more than one metre tall. A native of Mexico and Honduras this species grows on trees and rocks in the mountains usually in situations exposed to full sun for most of the day. Early collectors recorded that very exposed plants were common, having large reddish pseudobulbs and broad leathery leaves. *Laelia anceps* grows exceptionally well on a suitable tree but is more usually grown in pots or on slabs. An unheated glasshouse provides suitable shelter in temperate climates for this hardy species. Some attractive cultivars have been selected including 'Alba' – white flowers with a yellow patch on the labellum, 'Dawsonii' – large flowers with white sepals and petals and dark purple lines and suffusions on the labellum 'Chamberlains' – large flowers of a rich rosy purple colour and a prominent brightly coloured labellum.

Temp	Shade	In/on	Humidity	Air Flow
Cool	10-30%	Pot or Slab	50-70%	4

Laelia cinnabarina Bateman ex Lindley

(vermilion red)

This native of Brazil is popular with orchid growers for its wiry racemes of burnt orange to cinnabar-red flowers. Plant growth habit is neat and compact and a substantial flowering specimen does not occupy much space. Some forms of this species are disappointing in that the flowers are self-pollinating and short lived. Some variation exists in flower colour including some which have attractive yellow flowers. This rupicolous species grows almost exclusively on rocks in its native state.

Temp	Shade	In/on	Humidity	Air Flow
Int-Hot	10-30%	Small Pot	30-50%	5

Laelia cinnabarina

Laelia dayana

Laelia dayana Reichb.

(after John Day, 19th century English orchid grower who first flowered this species)

This Brazilian orchid has been included as a variety of *L. pumila*, and although plants of both species share the dwarf growth habit, the labellum of *L. dayana* has a much broader mid-lobe with dark purple margins and five to seven raised purple lines on a white central patch. This attractive species appears to be mainly a collectors item in Australia.

Temp	Shade	In/on	Humidity	Air Flow
Int-Hot	10-30%	Pot	30-50%	5

Laelia X gouldiana H. G. Reichb.

(after Jay Gould, American philanthropist)

This attractive orchid is apparently a natural hybrid between *L. anceps* and *L. autumnalis*. It is native to Mexico and is of rare natural occurrence but has become relatively common in cultivation. The colourful, widely opening flowers are well displayed on a tall flower stem (30-45 cm long).

Temp	Shade	In/on	Humidity	Air Flow
Cool-Int	10-30%	Pot	50-70%	5

Laelia X gouldiana

Laelia harpophylla H. G. Reichb.

(with leaves shaped like sickles)

Plants of this species have long, thin pseudobulbs each with a solitary, leathery leaf and the flowers are borne in crowded clusters on short stems. These are a colourful cinnabar-red except for the pale whitish labellum which usually curls back near the tip. A native of Brazil, this desirable orchid usually grows on trees rather than rocks, as do most other rupicolous species.

Temp	Shade	In/on	Humidity	Air Flow
Int-Hot	10-30%	Pot	30-50%	5

Laelia lucasiana Rolfe

(after M. Lucas)

The old pseudobulbs of this species turn bright orange-yellow and render plants fairly conspicuous on the exposed, lichen-clad rocks in the mountains of Brazil. A lovely miniature species, this orchid has a compact growth habit and usually bears a solitary flower on a relatively short, stiffly erect raceme. The colourful flowers have a contrasting bright yellow, wrinkled labellum. The roots of this orchid are very sensitive to lack of oxygen and plants must be grown in very coarse mixtures or attached to slabs of weathered hardwood or cork.

Temp	Shade	In/on	Humidity	Air Flow
Int	30-50%	Pot or Slab	30-50%	5

Laelia harpophylla

Laelia lucasiana

Laelia purpurata

Laelia pumila (Hook.) H. G. Reichb.

(dwarf, small)

A dwarf species with well-spaced psuedobulbs only 5-10 cm long, each topped by a solitary leaf. The disproportionately large flowers (to 10 cm across) are long lasting and release a pleasant fragrance into the atmosphere on warm days. With its dwarf habit, large colourful flowers and ease of culture, this species is deservedly popular with growers. A native of Brazil, it grows on trees on mountain slopes at 500 to 800 m altitude.

Temp	Shade	In/on	Humidity	Air Flow
Int-Hot	10-30%	Small Pot	30-50%	5

Laelia purpurata Lindley & Paxton

(purple)

A magnificent orchid which is native to Brazil and is the national flower of that country. It was first discovered by europeans in 1847 and quickly became a prized item in the orchid collections of Britain and Europe. Its flowers, which can be up to 20 cm across, are delightfully fragrant and last well. Because it is extremely variable in flower colour, many interesting selections and cultivars have been introduced into cultivation, some being prized subjects. In general this is an easy and rewarding orchid to grow.

Temp	Shade	In/on	Humidity	Air Flow
Int-Hot	30-50%	Pot	50-70%	5

Laelia pumila

Laelia xanthina Lindley

(yellow)

A Brazilian species which grows on trees at low to moderate elevations, usually in situations of good light. Some variants are disappointing because the flowers are self-pollinating and open poorly. Good clones have bright yellow, waxy flowers which open widely, last well and produce a fine display.

Temp	Shade	In/on	Humidity	Air Flow
Int-Hot	30-50%	Pot	50-70%	5

Laelia xanthina

LEPANTHOPSIS Ames

(from the Greek, *opsis*, resemblance; implying these orchids have a close resemblance to those in the genus *Lepanthes*)

A small genus of about 15 species found in Florida, Mexico, the West Indies, Central America and South America. Plants have a compact growth habit, lack pseudobulbs and with a single leaf terminal on each slender stem. The flowers are tiny but interesting in close up and are arranged in opposite pairs along a slender raceme.

Cultivation: These orchids have similar cultural requirements to species of *Pleurothallis*.

Lepanthopsis floripecten (H. G. Reichb.) Ames

(flowers arranged like a comb)

The tiny translucent pinkish or yellow flowers of this miniature orchid are carried in a most interesting manner being arranged in opposite pairs along an erect or arching, slender raceme. Plants have a compact habit and can be easily grown in a very small pot. A common species, it is native to many countries of Central America and South America at altitudes of about 1000 m.

Temp	Shade	In/on	Humidity	Air Flow
Int	30-50%	Pot	50-70%	5

Lepanthopsis floripecten

Ludisia discolor (flower)
Ludisia discolor

LUDISIA A. Rich

(derivation unknown)

A genus of a single species which is a very popular orchid in cultivation and for many years was incorrectly known in the genus *Haemaria*. It is still commonly labelled and exhibited by growers under the latter generic name, although *Ludisia* is gradually becoming more widely used.

Cultivation: Plants of this orchid grow readily in a shallow pot or saucer of well-drained, fibrous mix and lend themselves well to grouping together. They require intermediate growing conditions, high humidity and continual gentle air movement. While this species generally needs protected conditions, best colouration in the leaves is achieved in fairly bright, but diffuse, light; if light is excesssive, however the leaves will burn quickly. After flowering the plants have a dormant period when they should be kept dry. Continual protection is necessary against slugs and snails which relish the succulent tissues of this orchid.

Ludisia discolor (Ker-Gawler) A. Rich.

(of different colours)

This orchid is very popular with growers for both its handsome patterned foliage which is seen over much of the year and later in the season for its tall spikes of waxy white flowers. The leaves are generally of dark colouration but with distinctive reddish veins which almost seem to glow in certain light. The species is native to China, Thailand, Vietnam and Malaysia and grows in fairly open positions close to streams.

Temp	Shade	In/on	Humidity	Air Flow
Int	30-50%	Pot	50-70%	5

LYCASTE Lindley

(from the Greek word for a nymph; this genus was dedicated to the daughter of Priam, last King of Troy)

A genus of about 35 species found in the West Indies, Mexico, Central America and South America where they grow on trees or rocks from near sea level to high elevations in the mountains. All species have crowded pseudobulbs and large, thin-textured leaves which are shed as the growths mature. Large, waxy colourful flowers are borne singly on slender stems which emerge either as the pseudobulbs mature or with the onset of new growths. Flowering can be prolific with well-grown specimens producing exciting displays.

Cultivation: Some growers regard species of *Lycaste* as difficult subjects and while certain species do have specific requirements the commonly grown species succeed well in a mixed collection. With their fine roots drainage must be excellent and it is adviseable to greatly enlarge the drainage holes of a pot and fill the base with coarse material prior to potting. It is also best to underpot rather than overpot these orchids. Some growers use treefern fibre for potting, others prefer particles of bark, charcoal and leaf mould. Watering should be copious while the plants are in active growth and reduced as the leaves begin to fall and the plants enter their resting period. Water lodging in new growths is a common cause of rotting and susceptible growths should be dried after watering. Air movement is very important with these orchids as their large, soft leaves are highly susceptible to fungal diseases. They also need high humidity and although they will grow in fairly shady conditions, best flowering is achieved in good light.

Lycaste aromatica (Graham ex Hook.) Lindley

(aromatic, fragrant)

The delightful cinnamon aroma released by the flowers of this orchid, pervades the atmosphere of a glasshouse on warm days. A native of Mexico, Guatemala, Honduras and Nicaragua, it grows on trees mainly in the mountains at elevations of about 1500 m. Popular in cultivation it is one of the more easily grown members of the genus. Flowers arise in profusion from the base of a recently matured pseudobulb, usually coinciding with the onset of new growth.

Temp	Shade	In/on	Humidity	Air Flow
Int	30-50%	Pot	50-70%	5

Lycaste cruenta

Lycaste aromatica

Lycaste cruenta (Lindley) Lindley

(dull red, blood red, in reference to the red patch in the labellum)

Native of Mexico, Guatemala and El Salvador, this species grows in the mountains between 800 and 2000 m elevation. Popular with growers, its long-lasting, fragrant flowers arise from the base of a recently matured pseudobulb and coincide with the appearance of new growths. Similar in general appearance to *L. aromatica*, this species can be distinguished by the broadly rounded lateral lobes of the labellum.

Temp	Shade	In/on	Humidity	Air Flow
Int	30-50%	Pot	50-70%	5

Lycaste powellii Schltr.

(after R. W. Powell, original collector)

This species flowers while mature leaves are still present on the pseudobulbs. Attractively fragrant, the light green sepals are heavily blotched with red or brown and the short creamy petals are spotted and dotted with red while the labellum is white, sometimes spotted with red. Endemic to Panama, this species grows in valleys and ravines at low to intermediate elevations. It can be distinguished from all other species by the distinctly heart-shaped mid-lobe of the labellum.

Temp	Shade	In/on	Humidity	Air Flow
Hot	30-50%	Pot	50-70%	5

Lycaste powellii

Lycaste skinneri (Bateman & Lindley) Lindley

(after George Ure Skinner, original collector)

The national flower of Guatemala, this orchid is one of the most popular members of the genus being prized for its lovely pink, waxy flowers. In addition to Guatemala, the species also grows naturally in Mexico, El Salvador and Honduras. It is still sometimes sold as its synonym, *L. virginalis*. The flowers arise from the base of a mature pseudobulb while the leaves are still present and usually before the onset of new growth. A number of attractive cultivars have been named including a lovely one which has pure white flowers.

Temp	Shade	In/on	Humidity	Air Flow
Int	30-50%	Pot	50-70%	5

Lycaste skinneri

MASDEVALLIA Ruiz & Pavon

(after Dr Jose Masdevall, 18th century Spanish doctor and botanist)

Few orchid genera have captured the imagination of growers as has the genus *Masdevallia*. With a bewildering range to choose from including flowers in delicate pastel hues which often appear iridescent, extraordinary or even bizarre shapes, long spidery tips to the sepals, lilliputian species, these orchids have earned their place as a firm favourite with growers in many countries. The genus numbers more than 350 species and is distributed in Mexico, Central America and South America. The majority of species are found in high, mountainous regions where conditions are cool, moist and humid all year round. Under such conditions the plants have no requirement for moisture storage and storage organs such as pseudobulbs are unknown in the genus. All species have a compact growth habit and some are extremely free-flowering producing astonishing floral displays.

Cultivation: Species of *Masdevallia* have a reputation of being tricky to grow, however the realisation that most originate in cool, misty mountainous regions is a major beginning. In general they need cool, moist conditions with adequate air movement. Short periods of high temperature can be tolerated but prolonged exposure leads to their demise. Cool nights are very beneficial. Light requirements are generally not high but if the prevailing temperatures are low then plants will withstand bright light with a corresponding enhancement of flowering. Watering should be regular throughout the year with the compost just drying out before each watering. Humidity should be continually high together with good air movement. Masdevallias have very fine roots which rot quickly if the potting mix becomes clogged. For the same reason they should not be overpotted. Some growers repot annually concocting a mix from fern fibre, chopped sphagnum moss and fine particles of bark. Masdevallias are generally grown in pots but a few species which have pendulous inflorescences are best grown in small wooden baskets.

Masdevallia chimaera H. G. Reichb.

(the grotesque flowers bear a fanciful resemblance to the mythical monster, chimaera)

The wonderful flowers of this orchid have been compared with grotesque animals such as vampires. A native of Colombia, this species grows in dark, humid forests at altitudes between 1500 m and 2100 m on the trunks and lower branches of trees; less commonly on accumulations of fallen debris on the ground. The pendant flower spikes are produced in succession over many months, each bearing a single flower which lasts for a few days. First discovered in 1871, this species aroused considerable attention when plants were exhibited in Europe. Being extremely variable, numerous variants have been selected and named as cultivars or varieties. The first listing of these variations was made as early as 1875 by Gustav Wallis in the Gardeners Chronicle. *M. chimaera* remains a first favourite with orchid enthusiasts to this day.

Temp	Shade	In/on	Humidity	Air Flow
Cool	30-50%	Pot	70-90%	4-5

Masdevallia chimaera

Masdevallia coccinea

Masdevallia coccinea Lindley

(scarlet)

One of the most popular members of the genus, this species is prized among orchid growers for its displays of brilliantly coloured flowers, each being carried singly at the end of a long slender stem. The most striking colours range from scarlet to magenta or vermilion with a touch of purple, but pale yellow or white-flowered variants are also known. It is native to the Andes of Colombia growing typically on rocks at altitudes between 2000 m and 3000 m. Plants originating at higher altitudes generally have shorter leaves and larger flowers than those from lower down.

Temp	Shade	In/on	Humidity	Air Flow
Cool	30-50%	Pot	70-90%	5

Masdevallia infracta Lindley

(entire, unbroken)

This species is an outlier of the genus *Masdevallia* occurring as it does in the mountains of Brazil. Plants are characterised by extremely glossy leaves and have bell-shaped, multicoloured flowers in hues of yellow, cream and purple. With its compact growth habit and interesting flowers, this species is an adornment to any collection.

Temp	Shade	In/on	Humidity	Air Flow
Cool	30-50%	Pot	70-90%	5

Masdevallia infracta

Masdevallia tovarensis H. G. Reichb.

(from Colonia Tovar, the type locality)

Endemic to Venezuela, this species grows at moderate to high altitudes on rocks in protected situations where the humidity is constantly high. Plants form a dense clump of rigid, fleshy, deep green leaves which set off the intensely white flowers which are borne one to four at the end of sturdy, triangular stems which grow to about 15 cm long. Well grown plants flower heavily and produce an impressive display. This species is popular in cultivation.

Temp	Shade	In/on	Humidity	Air Flow
Cool	30-50%	Pot	50-70%	5

Masdevallia tovarensis

Maxillaria densa

MAXILLARIA Ruiz Lopez & Pavon

(from the Latin *maxilla*, jaw; the flowers of some species resemble the iaws of an insect)

Although this genus has captured the attention of many orchid growers, it has not achieved the heights of popularity reached by genera such as *Cattleya* and *Masdevallia*. This is perhaps surprising for it embraces a range of growth habits and most species have colourful, interesting flowers, some even being large and spectacular. With about 300 species, *Maxillaria* is a highly significant genus and provides an abundance of choice for growers who find these attractive orchids to their liking. Most species grow as epiphytes or lithophytes but a few are terrestrials. The genus ranges from southern Florida and Mexico to the West Indies and various countries in Central America and South America, extending well into the tropics of Argentina. While many species are found in the warmer forests of the lowlands, the genus reaches its greatest proliferation in the mountains especially the Andes in Peru and Brazil where most species occur.

All species of *Maxillaria* are evergreen, most with prominent clustered pseudobulbs but some species have a creeping rhizome and their pseudobulbs are much reduced in size or may be hidden under papery bracts. In this genus the flowers are always solitary on short, stiff stems and are sometimes produced in abundance and provide an attractive floral display.

Cultivation: With such a large genus it is difficult to generalise as to cultivation practices, however, it can be safely stated that the majority of species of *Maxillaria* tried have proved to be easy and rewarding subjects to grow. Those originating in the lowlands (especially from the tropics) require warmer conditions than do others from high altitudes. Clumping species are generally grown in pots whereas those with a creeping rhizome are best attached to slabs of treefern or cork. A few species which have a dwarf, compact habit are best accomodated in small pots. Those with a pendulous habit can be grown in hanging pots or baskets. Suitable potting materials include fern fibres and softwood bark, charcoal and sand. Conditions should be humid with an abundance of air movement. Watering should be regular except for a two or three week period of quiescence after the pseudobulbs have matured when plants are kept on the dry side. Excessive light causes leaching or burning of the foliage. Fertilisers are beneficial while plants are in growth.

Maxillaria densa Lindley

(dense, in reference to its dense flower clusters)

A common epiphyte with elongated, almost creeping rhizomes, usually with only the apical pseudobulbs having any leaves. Small flowers are borne singly at the end of short, slender stalks, the most interesting feature being that they arise in dense clusters of about twenty or thirty from each pseudobulb. The flowers are variable in colour from deep maroon or reddish brown to almost pure white. The species occurs in moist forests and ranges from low elevations to cloud forests at about 2,500 m altitude. It is found in Mexico, British Honduras, Guatemala and Honduras.

Temp	Shade	In/on	Humidity	Air Flow
Int	30-50%	Pot or Slab	50-70%	5

Maxillaria picta Hook.

(painted)

An attractive species from low to intermediate altitudes in Brazil. It is quite distinct from *M. porphorystele* with which it is often confused. Plants of *M. picta* are darker green with longer leaves and the bright, golden-yellow flowers are carried on longer stems (to 35 cm long). An excellent species for specimen culture.

Temp	Shade	In/on	Humidity	Air Flow
Int-Hot	10-30%	Pot	50-70%	4

Maxillaria porphyrostele H. G. Reichb.

(with a purple column)

An excellent beginners orchid which is so adaptable and tolerant of neglect that it is difficult to kill. Although commonly cultivated, especially in temperate regions, it is frequently confused with *M. picta*. From that species it differs by its paler, sulphur-yellow flowers borne on much shorter stems (to 8 cm long). Plants also tend to have shorter leaves which are usually yellowish green. It is native to the mountains of Brazil.

Temp	Shade	In/on	Humidity	Air Flow
Cool	10-30%	Pot, Basket or Slab	30-50%	4

Maxillaria picta

Maxillaria porphyrostele

Maxillaria tenuifolia Lindley

(with slender leaves)

The unusual, thick textured, dark red flowers of this orchid can hardly be described as showy, but they usually arouse comment and have a lovely perfume reminiscent of coconuts. They are fairly long lasting and contrast with the lustrous pseudobulbs and narrow, dark green leaves of the plants. Being easy to grow this species is fairly commonly cultivated. It is native to Mexico and some countries of Central America, growing in forests at low elevations although sometimes extending up to 1500 m altitude.

Temp	Shade	In/on	Humidity	Air Flow
Int	30-50%	Pot or Slab	50-70%	5

Maxillaria tenuifolia

Maxillaria variabilis Batem. ex Lindley

(variable)

As the specific name suggests, this species is somewhat variable especially in flower colour which ranges from wholly dark red to white or yellow with red markings and with or without a red labellum. Of very wide distribution this species is known from Mexico, the West Indies and many countries in Central America and South America. Plants grow variously as terrestrials or on rocks or trees up to 1900 m elevation. With its creeping rhizome, plants are best grown on slabs or on a treefern totem in a small pot of softwood bark or fern fibre.

Temp	Shade	In/on	Humidity	Air Flow
Int	30-50%	Pot or Slab	50-70%	5

Maxillaria variabilis

Miltonia flavescens

Miltonia flavescens Lindley

(yellowish)

Although its flowers are not highly ornamental, this species is well regarded by orchid growers for its adaptability and ease of culture. The star-like flowers last reasonably well, are fragrant and are produced regularly each year. Plants grow vigorously and are very suitable for specimen culture.

Temp	Shade	In/on	Humidity	Air Flow
Hot	10-30%	Pot, Basket or Slab	50-70%	5

Miltonia regnellii H. G. Reichb.

(after Dr. Regnell, original collector)

The large flat flowers of this attractive orchid are white with a prominent central patch on the labellum which is usually light mauve-pink, but is rich purple in one superior variant and wholly pink with darker veins in others. A native of Brazil, this species grows on trees and rocks in shady forests in mountainous regions between 800 m and 1400 m elevation. This species has been a favourite in cultivation almost since its introduction to european glasshouses in the 1850's.

Temp	Shade	In/on	Humidity	Air Flow
Cool-Int	30-50%	Pot or Slab	50-70%	5

MILTONIA Lindley

(after Viscount Milton, English patron of horticulture)

Although generally regarded as being a large genus, recent studies have split off segregate groups into other genera so that *Miltonia* now consists of only 4 species all of which are found in Brazil. They grow at low to intermediate altitudes and were once locally abundant but have now been reduced by clearing and over-collecting. Plants of true *Miltonia* species are yellowish green with a creeping rhizome, two leaves on each pseudobulb and prominent auricles on the column which is hollow in the front.

Cultivation: Species of *Miltonia* are easy to grow if provided with warm, humid conditions and an abundance of air movement. Pine bark and fir bark are suitable potting materials. With their fine roots, these orchids are very sensitive to stagnant conditions and excessive watering can quickly lead to rotting. Suitable potting materials include fern fibre (chunks or shredded), softwood bark, charcoal and coarse sand. Plants need regular watering throughout the year and bright, but diffuse, light for flowering.

Miltonia regnellii

Miltonia spectabilis Lindley

(showy, spectacular)

This orchid, the type species of the genus *Miltonia*, is native to Brazil and grows in moist, shady, mountainous forests. It is a very easy species to grow and lends itself well to specimen culture. Large plants flower freely and produce an impressive display. The flowers open very widely (almost flat) and are typically cream or white but may be sometimes tinged with mauve or rose pink.

Temp	Shade	In/on	Humidity	Air Flow
Int	30-50%	Pot or Slab	50-70%	5

Miltonia spectabilis

Miltonia spectabilis var. *moreliana*
Henfrey

(after M. Morel, 19th century French orchid grower)

A remarkable colour variant which is deservedly more popular in cultivation than the typical form. Its larger flowers have dark purplish black segments and the labellum is a lighter rose pink with darker veins. The flowers are most attractive and a well-grown plant produces an impressive display. Mostly plants of this orchid are grown in small pots but they also grow well on slabs of treefern.

Temp	Shade	In/on	Humidity	Air Flow
Int	30-50%	Pot or Slab	50-70%	5

Miltonia spectabilis var. moreliana

MILTONIOIDES Brieger & Luckel

(like the genus *Miltonia*)

A genus of 4 species of epiphytic orchids which have been variously included in the genera *Miltonia*, *Odontoglossum* and *Oncidium*. All are native to Mexico and Central America. **Cultivation**; These orchids have similar cultural requirements to species of *Miltonia* and *Oncidium*.

Miltonioides karwinski (Lindley) Brieger & Luckel

(after Count Karwinski)

This orchid has had a chequered history being placed in various genera including *Cyrtochilum*, *Odontoglossum*, *Miltonia* and *Oncidium* before its present status. A native of Mexico this species is rare in cultivation and is a true collector's item. Plants bear stiff, branched racemes up to 1 m long and have numerous, waxy colourful flowers each about 6 cm across. Colours include combinations of purple, yellow, brown, violet and white. Plants like cool, humid conditions with an abundance of air movement.

Temp	Shade	In/on	Humidity	Air Flow
Cool	30-50%	Pot	50-70%	5

Miltonioides karwinski

Miltoniopsis phalaenopsis

MILTONIOPSIS Godef.-Leb.

(resembling the genus *Miltonia*)

A genus of 5 species of orchids native to Costa Rica, Panama, Venezuela, Colombia and Ecuador. They grow on trees and rocks in cool, humid, shady forests. Until recently all species were generally accepted as members of the genus *Miltonia* but studies have confirmed their distinctiveness. The genus *Miltoniopsis* is not new, in fact it was established as early as 1889. Members can be recognised by their tufted habit, one leaf on each pseudobulb and no auricles on the column which is not hollow in front and has a prominent ridge leading onto the labellum. The plants of all species are grey or bluish green and contrast with the yellowish green of species of *Miltonia*.

Cultivation: These orchids have a reputation of being difficult to grow and are often considered mainly as subjects for the specialist grower. If their requirements are met, however they can be grown successfully in a mixed collection. Suitable conditions are intermediate temperatures, high humidity and an abundance of air movement. As a rule these orchids do not like high temperatures and may suffer a setback during hot summers. Pine bark and fir bark are suitable potting materials. Because they have fine roots these orchids are sensitive to stagnant conditions and excessive watering.

Miltoniopsis phalaenopsis (Nichols) Godef.-Leb.

(resembling a moth)
Pansy Orchid

A native of the mountains of Colombia, this species was first discovered in 1850 in abundance growing naturally on rocks in cloud forests where morning mists and afternoon showers are the norm. Originally classified as an *Odontoglossum* then a *Miltoniopsis*, it was then transferred to *Miltonia* where it resided for many years until its relatively recent transfer to the above genus. Its large flowers with the lovely waterfall markings on the labellum always attract attention, however the species has a reputation for being difficult to grow and is mainly found in the collections of specialists. Plants demand free unimpeded air movement with high humidity. They also prefer to be potbound and suffer a setback from repotting.

Temp	Shade	In/on	Humidity	Air Flow
Cool-Int	30-50%	Pot	50-70%	5

Miltoniopsis roezlii

Miltoniopsis roezlii (H. G. Reichb.) Godef.-Leb.

(after Benedict Roezl, original collector)

The first plants of this orchid were found on a log floating in the Dague River, Colombia in 1873. These plants were in flower and their distinctiveness was immediately recognised by the plant collector, B. Roezl. Subsequently the species has been found in Panama and Ecuador. Commonly it grows in shady positions on trees or rocks at 300 m-600 m elevation. Plants of this orchid are easily distinguished by their light bluish-green leaves.

Temp	Shade	In/on	Humidity	Air Flow
Int	30-50%	Pot	50-70%	5

MORMODES Lindley

(from the Greek, *mormo*, phantom; in allusion to the weird flowers of these orchids)

Members of this genus are commonly known as 'Goblin Orchids' because of the weird shape of their flowers. These are borne on arching racemes and have a curious twist in the column which causes the anther to face sideways. The genus is related to *Catasetum* and *Cycnoches* but the flowers are all bisexual. In all there are about 30 species of *Mormodes* and they are distributed in Mexico, Central America and South America. Most species grow as epiphytes but lithophytic and terrestrial habits are also known. While some are to be found in the lowlands, it seems that most grow at fairly high elevations in the mountains.

Cultivation: Species of *Mormodes* require excellent drainage and a coarse, fibrous mix is recommended. When in active growth the plants require an abundance of warmth, water and fertiliser with many revelling in higher light intensities than most orchids will tolerate. After the current growths have matured, plants lose their leaves and become dormant. During this phase flowering will occur and watering should be at a reduced level until the first signs of new growth.

Mormodes igneum Lindley & Paxton

(incandescent red)

A very common species native to Panama, Colombia and Costa Rica. Plants grow from the lowlands up to about 1300 m altitude and are usually found on exposed dead branches or on the tops of dead trees where they are subjected to full sun and wind. This species is remarkable for the variation in its flowers with plants growing in close proximity differing to an amazing degree in floral size, shape and particularly colour (yellow, green, brown, red-spotted or unspotted).

Temp	Shade	In/on	Humidity	Air Flow
Hot	10-30%	Pot	30-50%	5

Mormodes igneum

Mormodes maculatum

Mormodes maculatum (Kl.) L. O. Williams

(spotted)

A native of Mexico, this species is perhaps best known by the variant commonly called *unicolor* by orchid growers. In this the flowers are uniformly bright lemon yellow, whereas more commonly they are heavily spotted with dark red brown. This species is easily grown but the flowers emit a strong, off-putting, unpleasant odour.

Temp	Shade	In/on	Humidity	Air Flow
Int-Hot	10-30%	Pot	30-50%	5

NAGELIELLA L. O. Williams

(after Otto Nagel, Mexican orchid collector)

Perhaps better known to orchid growers by the synonym of *Hartwegia*, this is a small genus of 2 species both found in Central America. Each species is almost identical in growth habit but they are quite distinct florally (*N. angustifolia* lacks a spur on its labellum). Both grow as epiphytes at low to moderate altitudes.

Cultivation: Easy and rewarding, these small orchids deserve to become better known. Best results are achieved in a small pot of fern fibre in warm airy conditions. Good light is conducive to flowering and the leaves take on colourful hues. Racemes should never be removed from these orchids while still green as each is capable of sporadic flowering over many years.

Nageliella angustifolia

Nageliella angustifolia (Booth. ex Lindley) Ames & Correll

(with narrow leaves)

The leaves of this species are heavily marked with deep brown, reddish or purplish spots and produce an ornamental impact even when flowers are absent. Of natural occurrence in Guatemala, this species grows as an epiphyte in moist forests up to 2000 m altitude. The flowers are small but colourful and may be produced in succession over many months from a single raceme.

Temp	Shade	In/on	Humidity	Air Flow
Int	30-50%	Pot	50-70%	5

NEOFINETIA Hu

(originally named as *Finetia* but that name was taken up for another genus hence the prefix *neo* was used for distinguishing purposes)

A genus of a single species which is an attractive miniature deservedly popular with orchid growers and is of special importance in Japan.

Cultivation: Best grown in a small pot of well-drained fibrous mixture but can also be attached to a slab of cork or oak. This orchid is generally easy to grow if given high humidity with ample air movement. Watering should be regular throughout the year with a reduction in winter when growth slows.

Neofinetia falcata (Thunb.) Hu

(sickle-shaped)

For many years this dainty orchid was included in the genus *Angraecum* but with its singular flowers is best placed in its own genus. A dwarf species it is native to Japan and Korea where it grows as an epiphyte in cool forests. Plants have curved leaves and form neat clumps. The milky white flowers, borne on arching racemes, have a long spur and are pleasantly fragrant at night.

Temp	Shade	In/on	Humidity	Air Flow
Cool-Int	30-50%	Pot or Slab	50-70%	5

Neofinetia falcata

ONCIDIUM Sw.

(a diminutive of the Greek *oncos*, tumour or swelling; in reference to the wart-like calluses on the labella of these orchids)

A wonderful genus for orchidists since it contains sufficient numbers and in such a diversity of plant habit and floral features that a grower can spend a lifetime collecting all the variants and learning how to grow them. Numbering more than 750 species these orchids are distributed in south Florida, Mexico, the West Indies, Central America and South America. Ranging from the lowlands to high elevations in the mountains most species grow as epiphytes on trees and rocks with a few being terrestrials. Some distinct growth habits can be readily recognised within the genus. One group, the equitant species, lacks pseudobulbs and has iris-like growths of divergent leaves closely overlapping at the base. Another group, dubbed by growers as 'mule-ears' have large, thick leathery leaves. Still others have fleshy terete leaves. A distinctive group with mottled leaves has been separated as the genus *Psychopsis*.

Cultivation: Oncidiums are generally regarded as being an adaptable group of orchids but even so a consideration of their area of origin can help to avoid problems. Those species from high elevations in the Andes are cool growers whereas those originating in the lowlands require heat. Light and water requirements seem to be related and vary somewhat with the group involved. Those species which have tall, pear-shaped pseudobulbs (such as *O. sphacelatum* and *O. varicosum*) are in continual growth throughout the year and need to be kept continually moist and do not have a high light requirement. Terete-leaved species, those with very small pseudobulbs and those with nearly round pseudobulbs should be allowed to dry out between waterings (every two or three days). They will also tolerate bright light but if the leaves begin to turn red then reduce the light intensity somewhat. Equitant species are not as tolerant of dryness as the larger-leaved types. A short dry period after flowering will actually promote stronger growth.

Most species of *Oncidium* are grown in pots but some suffer badly from root rotting and are much better on slabs (e.g. *O. forbesii*, *O. gardneri*, *O. gravesianum*, *O. mashallianum*). Terete-leaved species are also better suited to slabs. Potting materials include bark flakes, charcoal and fern fibre. Under-potting is better than overpotting. Buoyant air movement is important for these orchids which as a general rule do not require high humidity.

Oncidium carthagenense (Jacq.) Sw.

(from Cartagena, North Colombia)

A very widely distributed species which is found naturally in the West Indies, south Florida, Mexico and various countries in Central America and South America as far south as Brazil. It grows on trees and rocks in humid forests from near the coast to mountains at about 1200 m altitude. Being widespread this species is variable especially in the size and colouration of the flowers. Many of these variants are grown in Australia.

Temp	Shade	In/on	Humidity	Air Flow
Int-Hot	10-30%	Pot or Slab	30-50%	5

Oncidium excavatum Lindley

(hollowed as though dug out, in reference to the pit at the base of the labellum)

A native of Ecuador and Peru, this orchid grows on trees and rocks in humid forests at intermediate elevations. It was first discovered in 1838 and quickly became popular with European orchid growers, especially those in England. Plants grow readily and flower quite freely. As with many species in this genus it is variable in flower size and colour; one particularly notable variant has wholly yellow petals.

Temp	Shade	In/on	Humidity	Air Flow
Int-Hot	30-50%	Pot or Slab	50-70%	5

Oncidium carthagenense

Oncidium excavatum

Oncidium forbesii

Oncidium forbesii Hook.

(after M. Forbes, gardener to Duke Bedford in whose collection this species first flowered)
Gold Lace Orchid

One of the handsomest members of the genus, this species is prized for its large, full flowers in which the petals bear a remarkable resemblance to the labellum. A native of Brazil, it grows on trees in mountainous regions and was first collected in 1837. Plants are generally easy to grow and a large plant with many racemes of flowers produces a very impressive display.

Temp	Shade	In/on	Humidity	Air Flow
Int-Hot	30-50%	Pot or Slab	50-70%	5

Oncidium incurvum G. Barker

(the petals curve inwards when the flowers first open)

A Mexican species which grows on trees and rocks at fairly high elevations (1200-1600 m). Its attractively marked flowers (about 2.5 cm across) are borne on wiry, arching panicles which may reach more than one and a half metres in length. The flowers are long-lasting and fragrant.

Temp	Shade	In/on	Humidity	Air Flow
Int	10-30%	Pot, Basket or Slab	30-50%	5

Oncidium lietzei Regel

(after M. Lietze, original collector)

A beautiful species from Brazil which appears to be mainly a collector's item. The flowers are unusual in that the lateral sepals are united to form a single unit similar in shape to the dorsal sepal but with a shallow notch at the apex. They are colourful, long-lasting and produced on slender, arching racemes or panicles to about one metre long.

Temp	Shade	In/on	Humidity	Air Flow
Hot	10-30%	Pot, Basket or Slab	30-50%	5

Oncidium incurvum

Oncidium maculatum

Oncidium micropogon

Oncidium maculatum Lindley

(spotted)

A native of Mexico, Guatemala and Honduras, *O. maculatum* grows on trees in humid forests up to about 2000 m altitude. Plants in general resemble those of *O. tigrinum* and the flowers are also similar but are duller with a paler, much less flared labellum. The long-lasting flowers are pleasantly fragrant.

Temp	Shade	In/on	Humidity	Air Flow
Int	10-30%	Pot, Basket or Slab	30-50%	5

Oncidium micropogon H. G. Reichb.

(with a small beard)

This species is one of about three Oncidiums which have fairly small flowers and an unusual labellum in which the yellow lateral lobes spread widely like wings or an extra pair of petals; this gives the flower a crowded appearance. *O. micropogon* can be distinguished from the others by its kidney-shaped mid-lobe. This species is native to Brazil where it grows on trees in fairly sparse forests.

Temp	Shade	In/on	Humidity	Air Flow
Int-Hot	10-30%	Pot, Basket or Slab	50-70%	5

Oncidium lietzei

Oncidium ornithorhynchum

Oncidium ornithorhynchum Kunth

(like the beak of a bird)

A beautiful species which produces highly ornamental displays of rose-pink flowers on arching or pendulous panicles each about 60 cm long. The long-lived, fragrant flowers, each about 2 cm long, seem to dance in the slightest breeze. Plants are floriferous and commonly produce two inflorescences from each recently mature pseudobulb. The species is native to Mexico, Guatemala, El Salvador and Costa Rica and grows on trees in humid mountainous forests at 1000-1500 m elevation.

Temp	Shade	In/on	Humidity	Air Flow
Int	10-30%	Pot, Basket or Slab	30-50%	5

Oncidium sphacelatum Lindley

(with brown or blackish speckling)

One of the commonest orchids grown and valued for its adaptability, ease of culture, showy floral displays and long flowering period. In tropical and subtropical regions plants are a familiar sight attached to garden trees and large flowering specimens are most noticeable. Common in nature, it grows on trees or rocks in lowland forests up to 800 m altitude and is native to Mexico, Central America and South America.

Temp	Shade	In/on	Humidity	Air Flow
Cool-Int	10-30%	Pot, Basket or Slab	30-50%	5

Oncidium ornithorhynchum (flower)

Oncidium sphacelatum

Oncidium tigrinum Llave & Lex.

(tiger-like, from the floral markings)

This popular orchid occurs in forests at low to intermediate elevations in Mexico, Guatemala and Honduras. Plants have prominent leaves on short, rounded pseudobulbs and bear their waxy, long-lasting flowers on tall, branched panicles. The prominent, large, yellow, flared labellum is most noticeable. Large plants produce an impressive floral display.

Temp	Shade	In/on	Humidity	Air Flow
Int	10-30%	Pot, Basket or Slab	30-50%	5

Oncidium varicosum Lindley

(swollen or dilated with veins)

An orchid which is commonly cultivated in many countries of the world and which is very adaptable to a wide range of growing conditions. In the tropics and subtropics plants are commonly seen attached to garden trees whereas in colder climates the protection of a glasshouse is necessary. With its profuse displays of cheery yellow flowers this orchid has become a firm favourite with orchid growers and is even familiar to the general gardening public. Of natural occurrence in Brazil, *O. varicosum* grows on trees in fairly bright light. The illustration is of the variety *rogersii* H. G. Reichb.; a very popular variant with a large labellum having three prominent apical clefts.

Temp	Shade	In/on	Humidity	Air Flow
Int	10-30%	Pot, Basket or Slab	30-50%	5

Oncidium varicosum

Oncidium tigrinum

PAPHIOPEDILUM Pfitzer

(from the Greek, *Paphos*, a name for Aphrodite, the goddess of love; p*edilon*, a slipper; literally meaning the slipper of Aphrodite)
Slipper Orchids

The common name of these orchids arises from the shape of their deeply pouched labellum. As a group they are so popular in cultivation that a number of species have nearly become extinct as a direct result of over-exploitation of wild populations by unscrupulous collectors responding to high prices and a nearly unlimited market. One result of this intense demand for plants has been a proliferation of names applying to new species and minor variants which are best treated as cultivars. The most recent assessment of this important genus puts it at 60 species, a number of varieties and numerous horticultural variants. The genus ranges from India to southern China, South-east Asia, Malaysia, the Philippines, Indonesia, New Guinea and the Solomon Islands. The majority grow as terrestrials in pockets of debris in rocks, in sandy soils, among tree roots, with a few lithophytic species attaching their roots firmly to rocks and about 5 species of epiphytes. A number grow on formations of limestone. While most species favour shady forests where they might receive dappled sunlight, a couple of species grow in sunny locations. Despite an absence of pseudobulbs, these orchids are quite able to withstand fairly long dry periods; most species grow naturally in areas where the prevailing climates are strongly seasonal with a long dry season usually in winter. Adaptations include long roots and tough, leathery leaves.

Hybridisation of slipper orchids has resulted in the registration of countless hybrid cultivars. The progeny of some differ so much from the original species that it is often difficult to recognise features which they could have contributed. While the hybrids largely dominate modern collections there are many growers who specialise in the species.

Cultivation: Species of *Paphiopedilum* have been cultivated for more than one hundred and fifty years and their basic cultural requirements are well established. The majority are grown in pots and plants should be underpotted rather than being overpotted. Potting mixes are made up of softwood bark, treefern fibre, charcoal, scoria, perlite, chopped sphagnum moss and limestone chips. A wide variety of mixes are used by growers, the particle size of the materials used being on average 5-10 mm across. Generally these orchids are tolerant of low light and although an enhancement of flowering may be noticed in bright conditions this may also result in bleached, yellowish leaves and washed out flowers on shorter stalks. Adequate shading over summer is of major importance for these orchids. Summer temperatures should be on the cool side as extremes of heat are very detrimental to their health. The humid air from evaporative coolers is appreciated and good ventilation should create a gentle, unimpeded flow of air. If the glasshouse has to be shut to retain humidity then fans should be employed to maintain air movement. Misting is a beneficial practice to keep plants cool in summer. Watering should be daily in spring and summer, tapering off in autumn and in winter once or twice a week is usually sufficient. More plants are lost from overwatering than underwatering and remember that these plants are quite tolerant of dry periods especially those with well established roots.

Paphiopedilum bellatulum (H. G. Reichb) Stein
(neat and beautiful)

A very neat species which of interest for its heavily mottled leaves and relatively large flowers (about 7 cm across) which because of their short stems almost seem to squat in the centre of the growths. The flowers are mostly white or cream and with large spots and blotches of dark maroon. Common in Thailand and Burma, this species grows in shady humid forests on outcrops of limestone at elevations of between 1000 m and 1500 m.

Temp	Shade	In/on	Humidity	Air Flow
Int	30-50%	Pot	50-70%	5

Paphiopedilum bellatulum

Paphiopedilum ciliolare (H. G. Reichb.) Stein
(finely fringed)

First discovered in 1882, this elegant orchid is distributed on many islands of the Philippines where it grows in shady forests at low to intermediate elevations. It is very closely related to *P. superbiens* and is treated as a subspecies of that taxon by some authors. Two useful features of distinction are the presence of numerous raised spots on the petals of *P. ciliolare* and two blunt marginal teeth on the apical side of the staminode of *P. superbiens*.

Temp	Shade	In/on	Humidity	Air Flow
Hot	30-50%	Pot	50-70%	5

Paphiopedilum ciliolare

Paphiopedilum concolor (Lindley) Pfitzer
(of uniform colouration)

One of the most delightful species in the genus, this orchid adapts readily to cultivation and flowers freely. Its pale yellow flowrs are dotted with minute purple spots and are usually borne on short stems close to the heavily mottled, dark green leaves. Widely distributed in southern China, southern Burma, Thailand, Laos, Cambodia and Vietnam, it grows in shady, humid forests at low to intermediate altitudes. Plants usually grow on limestone outcrops but have also been found in accumulations of sand.

Temp	Shade	In/on	Humidity	Air Flow
Int	30-50%	Pot	50-70%	5

Paphiopedilum concolor

Paphiopedilum druryi (Bedd.) Stein

(after Colonel H. Drury)

Although superficially similar to *P. insigne*, this species is readily recognised by its broader dorsal sepal which is distinctly ovate in shape. The handsome flowers of this easily grown species are well displayed, colourful, and long lasting. Unfortunately this orchid appears to be extinct in the wild due to destruction of its habitat by burning and overcollecting. It was only known from restricted localities in the Travancore Hills, India. Plants are well established in cultivation.

Temp	Shade	In/on	Humidity	Air Flow
Int	30-50%	Pot	50-70%	5

Paphiopedilum druryi

Paphiopedilum fairrieanum (Lindley) Stein

(after R. Fairrie, who first exhibited this species in 1857 in England)

A very distinctive species readily recognised by its relatively large, boldly striped dorsal sepal with strongly wavy margins and its strongly curved petals. Restricted in its natural distribution to the mountains of northern India at elevations between 1400 m and 2200 m, this species has unfortunately been so overcollected that it is now in danger of extinction in the wild. It grows on ledges and rocky slopes in shady forests on both acidic and alkaline rock formations.

Temp	Shade	In/on	Humidity	Air Flow
Int	30-50%	Pot	50-70%	5

Paphiopedilum fairrieanum

Paphiopedilum gratrixianum (Masters) Guillaumin

(after M. Gratrix)

Although of similar general appearance to *P. insigne*, this species has smaller flowers with shorter petals and a distinctly elliptical dorsal sepal. The prominent purple spots on the base of the leaves are also a useful guide. This species is known from Laos and believed to also occur in Vietnam, but details of its habitat are lacking.

Temp	Shade	In/on	Humidity	Air Flow
Hot	30-50%	Pot	50-70%	5

Paphiopedilum gratrixianum

Paphiopedilum hirsutissimum

Paphiopedilum haynaldianum (H. G. Reichb.) Stein

(after Archbishop Haynald, 19th century Polish amateur botanist)

In its natural state on the islands of Luzon and Negros in the Philippines, this species grows in humid forests from sea level to about 1400 m elevation. Most commonly it grows on rocks and boulders (both of acidic and alkaline formations) but sometimes it occurs as an epiphyte on trees. With its attractive, long-lasting flowers on long, arching racemes this species is a firm favourite with growers.

Temp	Shade	In/on	Humidity	Air Flow
Int	30-50%	Pot	50-70%	5

Paphiopedilum hirsutissimum (Lindley ex Hook) Stein

(very hairy)

With its very hairy flower stalks and prominently spathulate, purplish petals (strongly wrinkled and wavy at the base), this distinctive species is unlikely to be confused with any other. Native to the hills of northern India, it grows on rocks and trees at elevations between 700 m and 1200 m. Unfortunately it is now rare in nature due to overcollecting. This species grows and flowers well in cultivation.

Temp	Shade	In/on	Humidity	Air Flow
Int	30-50%	Pot	50-70%	5

Paphiopedilum insigne (Wall. ex Lindley) Pfitzer

(remarkable)

A long time favourite not only with orchidists but with gardeners in general. This species is valued for its hardiness, ease of culture and the displays of handsome, long-lasting flowers which it produces so readily. Plants grow steadily and soon build up into a potful. In its native state of north-eastern India, it grows in shady forests and near waterfalls on limestone outcrops between 1000 m and 1500 m elevation. Being rather variable a number of varieties have been described mainly based on flower size and colour. These variants are best treated as cultivars.

Temp	Shade	In/on	Humidity	Air Flow
Cool	30-50%	Pot	50-70%	5

Paphiopedilum haynaldianum

Paphiopedilum insigne

Paphiopedilum lowii (Lindley) Stein

(after Hugh Low & sons, original discoverers)

A true epiphyte, this handsome species provides impressive floral displays with its tall, multi-flowered racemes. Similar in general appearance to *P. haynaldianum* but distinguished by the absence of spots on the dorsal sepal and much narrower petals. Native to Malaysia, Indonesia and Borneo, this species grows on trees and rocks on hills and mountains at elevations between 250 m and 1600 m.

Temp	Shade	In/on	Humidity	Air Flow
Int-Hot	30-50%	Pot	50-70%	5

Paphiopedilum malipoense Chen & Tsi

(from the type locality, Malipo, China)

One of a group of recent discoveries in China, this species was described in 1984. With its attractively mottled leaves, neat green flowers and the intriguing inflated labellum, this species has tremendous appeal to growers. Details of its natural habitat are lacking but it is of restricted occurrence in the Yunnan Province of southern China in montane forest at about 1500 m elevation.

Temp	Shade	In/on	Humidity	Air Flow
Int	30-50%	Pot	50-70%	5

Paphiopedilum malipoense

Paphiopedilum lowii

Paphiopedilum micranthum T. Tang & Wang
(with small flowers)

One of the most attractive of the Chinese species, *P. micranthum* grows as a terrestrial on limestone hills in the Yunnan Province, near the border with Vietnam. The original description was based on a very small-flowered specimen but this is atypical of the species. Although only described in 1951, this species is now quite commonly grown and is regularly exhibited at shows. These collections have depleted natural populations to the extent that it is now believed to be endangered in the wild.

Temp	Shade	In/on	Humidity	Air Flow
Int-Hot	30-50%	Pot	50-70%	5

Paphiopedilum micranthum

Paphiopedilum niveum (H. G. Reichb.) Stein
(snow white)

A delightful species similar in some respects to *P. concolor* but with intense white flowers of very different shape on taller stems. In nature this species has a fairly restricted distribution in northern Malaysia and southern Thailand and sadly it has been much depleted by overcollecting. It grows as a terrestrial on outcrops of limestone in shady forests at low elevations.

Temp	Shade	In/on	Humidity	Air Flow
Int-Hot	30-50%	Pot	50-70%	5

Paphiopedilum niveum

Paphiopedilum parishii (H. G. Reichb.) Stein
(after Rev. Charles Parish, original collector while a missionary in Burma)

The green and blackish flowers of this species are most distinctive and when combined with its long twisted petals, make this orchid one of the most impressive members of the genus. Of natural occurrence in Burma, Thailand and southern China where it grows at moderate to high altitudes (1200-2200 m), this handsome orchid is still fairly rare in cultivation. In nature this orchid grows on rocks and boulders, less commonly on trees, in dense shade.

Temp	Shade	In/on	Humidity	Air Flow
Int	30-50%	Pot	50-70%	5

Paphiopedilum parishii

Paphiopedilum philippinense (H. G. Reichb.) Stein

(from the Philippines)

A hardy sun-loving species which grows naturally on limestone outcrops, boulders and cliffs from sea level to about 500 m elevation. Plants are widely distributed throughout many islands of the Philippines and it has also been recorded from an island to the north of Sabah. With its colourful flowers having impressively long, drooping, twisted petals, and being carried on strong erect racemes, this species is much sought after by orchid growers.

Temp	Shade	In/on	Humidity	Air Flow
Hot	30-50%	Pot	50-70%	5

Paphiopedilum primulinum

Paphiopedilum spicerianum (H. G. Reichb. ex Masters & T. Moore)

(after Herbert Spicer, who flowered this species in 1875)

The broad, white dorsal sepal of this species with its prominent red central stripe provides a ready means for its identification. Native to north-eastern India and north-western Burma, this attractive species grows on limestone escarpments and cliffs between 300 m and 1300 m elevation. Plants occur in shady situations with their leaves hanging loosely. Mists are frequent in the area and the limestone is moistened by seepage.

Temp	Shade	In/on	Humidity	Air Flow
Int	30-50%	Pot	50-70%	5

Paphiopedilum philippinense

Paphiopedilum primulinum M. Wood & Taylor

(primrose-like, probably a reference to the flower colour)

A recently discovered species which was described in 1973 from plants discovered in northern Sumatra. Here it grows in accumulations of litter on limestone hills at about 500 m altitude. Plants grow in shady positions in stunted forest. Prized for its beautiful yellow flowers, this species appears to be of restricted natural occurrence and is rare in cultivation.

Temp	Shade	In/on	Humidity	Air Flow
Hot	30-50%	Pot	50-70%	5

Paphiopedilum spicerianum

Paphiopedilum sukhakulii Schoser & Senghas

(after Prason Sukhakul, original collector)

Although very common in cultivation, the impressive flowers of this species never fail to create interest. Described as recently as 1967, this species was first discovered in a consignment of plants received in West Germany and originating in Thailand. By tracing the consignment the new species was discovered on a mountain in north-eastern Thailand at about 1000 m elevation. This is the only known locality for the species and it is now nearly extinct in the wild due to overcollecting.

Temp	Shade	In/on	Humidity	Air Flow
Int	30-50%	Pot	50-70%	5

Paphiopedilum sukhakulii

Paphiopedilum superbiens (H. G. Reichb.) Stein

(magnificent, proud)

Mountains on the Indonesian island of Sumatra are the home of this lovely orchid. It grows as a terrestrial on steep slopes at altitudes between 800 m and 1300 m in shady forests. Plants grow well in cultivation and are valued for their elegant, colourful flowers. Plants grown as *P. curtisii* are regarded by authorities as being synonymous with *P. superbiens* but many growers maintain that these two are distinct.

Temp	Shade	In/on	Humidity	Air Flow
Int	30-50%	Pot	50-70%	5

Paphiopedilum venustum (Wall.) Pfitzer

(pleasant, attractive)

This species was once common in the hills and mountains of northern India but clearing and depradations by collectors have reduced its levels to the verge of extinction. Originally discovered in 1816 and introduced to cultivation soon after, this species grows readily and flowers well. In nature it grows on shaded slopes above streams and in accumulations of litter in shady, humid forests up to 1300 m altitude.

Temp	Shade	In/on	Humidity	Air Flow
Int	30-50%	Pot	50-70%	5

Paphiopedilum superbiens

Paphiopedilum venustum

Paphiopedilum villosum (Lindley) Stein

(with shaggy hairs)

An old time favourite, this orchid is popular in cultivation because it is so easy to grow and flower. Of natural occurrence in northern India, Burma and Thailand, this species grows in mountainous habitats between 1000 m and 2000 m elevation. Surprisingly this species commonly grows as an epiphyte on mossy trees, less commonly on outcrops of rock.

Temp	Shade	In/on	Humidity	Air Flow
Cool	30-50%	Pot	30-70%	5

Paphiopedilum villosum

PERISTERIA Hook.

(from the Greek, *peristeria*, a dove; the column and labellum of *P. elata* look remarkably like a dove at rest)

A genus of about 6 species of orchids originating in Central America and South America. They grow as terrestrials or epiphytes and are somewhat diverse in their growth habits and flowers. Only one species, *P. elata* is commonly grown in Australia.

Cultivation: Species of *Peristeria* are deciduous and have a distinct dormant period during which they should be kept dry. Drainage of the potting mix must be excellent. Species with a terrestrial habit are potted into fern fibre, coarse sand and leaf mould whereas the epiphytes will grow in fern fibre and bark. When in active growth the plants should be watered at least once daily. Suitable conditions include warmth, humidity and bouyant air movement. All species need bright light to promote flowering.

Peristeria elata Hook.

(stately, tall; in reference to the racemes)

This species is the national flower of Panama and it also occurs in Costa Rica, Venezuela and Colombia. It grows as a terrestrial in loamy soil and humus pockets among rocks. The plants have a general similarity to many species of *Phaius* but the cupped, white waxy flowers are distinctive. The inflorescence arises with the new growth but develops slowly so that the growth is half mature before the flowers open. The leaves are shed as the pseudobulbs swell and mature and the plants are then dormant for some weeks. This orchid makes an admirable garden plant in tropical areas, being well suited to culture under trees with a light canopy.

Temp	Shade	In/on	Humidity	Air Flow
Hot	30-50%	Pot	50-70%	5

Peristeria elata

PESCATOREA H. G. Reichb.

(after M. Pescatore, 19th century French orchidologist)

A genus of about 12 species all of which are epiphytes in Central America and South America. At one time all were included in the genus *Zygopetalum* but the plants lack pseudobulbs and the flowers are borne singly on short stems. Most species have a growth habit very similar to that of *Huntleya meleagris*. Their flowers are extremely showy and while they are mainly for the collector, all species deserve to become better known.

Cultivation: These orchids can be difficult to grow and since they lack pseudobulbs they should never be allowed to dry out. All respond to cool to warm, moist conditions with plenty of air movement. Bright light is not essential for flowering. A small pot of fern fibre is suitable for growth.

Pescatorea cerina

Pescatorea cerina (Lindley & Paxton) H. G. Reichb.

(wax-coloured)

This attractive orchid grows as an epiphyte in shady situations in moist to wet highland forests of Costa Rica and Panama. The plants lack pseudobulbs and have a fan of arching, leathery, pleated leaves and large, showy, widely opening flowers well displayed on the end of short stems. These waxy flowers last well and have a heavy, spicy perfume.

Temp	Shade	In/on	Humidity	Air Flow
Int-Hot	50-70%	Pot	50-70%	5

Pescatorea dayana H. G. Reichb.

(after John Day, pioneer English orchid grower)

A native of Colombia, this species is similar in many respects to *P. cerina* but is more vigorous with larger flowers and a colourful labellum flushed with violet or magenta. Usually the flowers have greenish tips on the sepals but occasionally these are red or violet.

Temp	Shade	In/on	Humidity	Air Flow
Int-Hot	30-50%	Pot	50-70%	5

Pescatorea dayana

PHALAENOPSIS Blume

(from the Greek, *phalaina*, a moth; opsis, resemblance, appearance; in reference to the white-flowered species resembling large moths)

The genus *Phalaenopsis* consists of about 35 species which are found in the tropical parts of India, China, Malaysia, South-east Asia, Indonesia, the Philippines, New Guinea and Australia. With their long arching racemes of showy, widely-opening flowers they are amongst the most beautiful of all orchids and are deservedly popular in cultivation. All species are epiphytes and the majority originate from tropical lowlands, growing on trees close to streams, particularly those adjacent to rapids and waterfalls. Plants are monopodial in their growth habit and may flower over many months of the year.

Cultivation: *Phalaenopsis* are easy orchids to grow if given their basic requirements, however they also die quickly if they are unhappy with their compost or surroundings. In general they are shade lovers and will grow and flower well under conditions of low light. Conditions must be warm to hot, with high humidity and regular, gentle air movement. If kept sufficiently warm, growth is continuous throughout the year but if temperatures dip and photoperiod reduces sufficiently in winter then growth slows or may even cease. Species of *Phalaenopsis* have no capacity for water storage and hence must never be allowed to dry out severely. They are heavy feeders and respond well to applications of slow release fertilisers and liquid preparations. They are commonly grown in pots but some growers prefer slatted baskets and plants can also be attached to slabs of treefern or oak. Many different materials can be used successfully for potting including charcoal, fern fibre and softwood bark. Drainage must be excellent although once these orchids become established most roots grow through the air rather than into the potting mix. After flowering, the racemes of these orchids should be cut back to the node which supported the first flower and often secondary racemes will develop from buds lower down.

Phalaenopsis amabilis (L.) Blume

(lovely)
White Moth Orchid

This, the type species of the genus, is one of the most popular in cultivation and has been widely used in hybridisation. It is said that the name *Phalaenopsis* was coined when early botanists, on observing a flowering plant through field glasses, mistook the flowers for a group of moths. A native of Indonesia, Borneo and the Celebes, it grows on trees often close to the coast but also in rainforests further inland. Well grown plants are rarely out of flower and the arching sprays of large white flowers produce an impressive display.

Temp	Shade	In/on	Humidity	Air Flow
Hot	50-70%	Pot, Basket or Slab	70-90%	5

Phalaenopsis amboinensis
Phalaenopsis amabilis

Phalaenopsis amboinensis J. J. Smith

(from Ambon)

An easy growing, free-flowering species which is becoming popular with orchid growers and is also being used in breeding programs. It occurs naturally in the Moluccas and also Sulawesi growing on trees in warm, humid forests. The colourful, long-lasting flowers are heavily barred with cinnamon or tan and although only borne on short racemes are noticeable and showy.

Temp	Shade	In/on	Humidity	Air Flow
Hot	50-70%	Pot, Basket or Slab	70-90%	5

Phalaenopsis equestris H. G. Reichb.

(pertaining to horses)

A native of the Philippines, *P. equestris* grows in warm, shady, humid forests of the lowlands. Plants have slightly mottled leathery leaves and the arching racemes last for a couple of years, not branching until the second year. The flowers open in succession and only a few are open at any one time. They are hardly spectacular but are colourful and interesting when examined closely.

Temp	Shade	In/on	Humidity	Air Flow
Hot	50-70%	Pot, Basket or Slab	70-90%	5

Phalaenopsis lueddemanniana H. G. Reichb.

(after M. Luddemann, 19th century French orchid grower)

The colourful flowers of this dainty species are usually heavily barred with red or brown but in some variants have fine bands and lines like hieroglyphics. The flowers are also waxy or even glossy, long-lasting and noticeably fragrant on warm days. A native of the Philippines, this species is easily grown and flowers freely.

Temp	Shade	In/on	Humidity	Air Flow
Hot	50-70%	Pot, Basket or Slab	70-90%	5

Phalaenopsis pulchra

Phalaenopsis pulchra (H. G. Reichb.) H. Sweet

(beautiful)

Some botanists treat this orchid as a mere colour form of *P. lueddemanniana*. Both occur naturally in the Philippines, however, *P. pulchra* can be immediately distinguished by its striking flowers with narrower segments which are a uniform pink, magenta or amethyst purple. Although only about 4 cm across, these flowers are noticeable not only for their colour but also their glossy or waxy lustre.

Temp	Shade	In/on	Humidity	Air Flow
Hot	50-70%	Pot, Basket or Slab	70-90%	5

Phalaenopsis schilleriana H. G. Reichb.

(after Count von Schiller, 19th century German orchid grower in whose collection the species first flowered)
Pink Moth Orchid

A superb orchid which produces breathtaking displays of delicate pink flowers which seem to dance on the wiry, arching, branched racemes. The plants themselves are handsome with the dark green leaves being mottled and marbled with grey and are usually purplish beneath. Although the flowers open bright pink they often fade quickly especially if the prevailing temperatures are high. A native to the Philippine islands of Luzon and Mindanao, this species is a common epiphyte in the mountains at 800-1200 m altitude.

Temp	Shade	In/on	Humidity	Air Flow
Hot	50-70%	Pot, Basket or Slab	70-90%	5

Phalaenopsis lueddemanniana

Phalaenopsis schilleriana

Phalaenopsis equestris

Physosiphon tubatus

PHYSOSIPHON Lindley

(from the Greek *physa*, bellows; *siphon*, a tube; in reference to the inflated tube formed by the united sepals)

A genus of about 20 species of miniature orchids distributed in Mexico, Central America and South America. They grow as epiphytes on mossy trees and rocks in conditions which are moist and humid for most of the year. Elevations where they grow range from sea level to about 3000 m. The plants are generally small and compact with elongated racemes of small flowers in which the sepals are united in the basal half to form a tubular structure. All species lack pseudobulbs.

Cultivation: Species of *Physosiphon* have similar requirements to the closely related genera *Pleurothallis* and *Restrepia*. Most species are grown in small pots of finely particled materials including pine bark, charcoal, fern fibres and chopped sphagnum moss. Drainage must be excellent and repotting should take place immediately clogging of the mixture is noticed. Species originating in the lowlands require warmth whereas high altitude species need cool temperatures. Watering should be daily in summer and two or three times a week in winter. High humidity is also necessary combined with gentle air movement.

Physosiphon tubatus Rendle

(trumpet-shaped)

An interesting orchid which has clustered slender stems, each topped by a single leaf, and slender arching racemes (sometimes to 30 cm long) which have small, colourful flowers scattered along their length. These flowers vary tremendously in colour, (green to greenish yellow or bright red) and may be erect or drooping. A very rewarding little orchid which grows easily and flowers freely. It is native to Mexico, Guatemala, El Salvador, Honduras and Nicaragua and grows on trees in the mountains up to 3500 m elevation.

Temp	Shade	In/on	Humidity	Air Flow
Int	30-50%	Pot or Slab	50-70%	5

Physosiphon tubatus (flower)

PLEIONE D. Don

(named for *Pleione*, mother of the seven Pleiades)

A group of cool to cold growing orchids which tend to be neglected by growers possibly because they flower while leafless or perhaps they lack the challenge of some of their more flamboyant tropical cousins. They certainly do not lack beauty for their flowers are extremely ornamental and colourful. Plants are easy to grow although it is true that they do not succeed in hot climates, in fact most success is achieved in temperate regions and many species will tolerate winter temperatures as low as –5°C.

In all there are about fourteen species of *Pleione* and they are distributed from northern India and China, to Burma, northern Thailand, Laos and Taiwan. Plants of mountainous regions, they occur naturally at elevations from 600 m to about 4,200 m. Plants have clustered pseudobulbs and thin pleated leaves similar to species of *Coelogyne*. These leaves are shed at the end of the growing season and plants are dormant during the dry, cold winter. Flowering of most species takes place in autumn but some flower in spring. These orchids grow as terrestrials in gravelly soil, in pads of moss, on rock outcrops or as epiphytes on mossy tree trunks. Some species exhibit great variation in flower colour and their is good scope for selection of unusual strains and hybridisation within species. *Pleione* species reproduce freely by the production of small bulbils from the top of pseudobulbs. These can be removed, potted and will mature to flowering size plant in two to four years.

Cultivation: Successful growers of these orchids in southern Australia house them in unheated glasshouses, bushhouses or even cold frames. Some species are suitable for rockery culture but must be continually protected from slugs and snails. Generally these orchids are grown in squat pots and are repotted annually when dormant. At this time leaf sheaths and old flower stems are cut off and the roots trimmed back to within 2 cm or 3 cm of the pseudobulbs.

Potting materials include

softwood bark	6 parts
chopped sphagnum moss	3 parts
leaf mould	2 parts
perlite	1 part
charcoal	1 part

Plants should be kept on the dry side over winter and regular watering begins with appearance of new shoots and roots. Fertilising is a benefit during the period of major growth. Ventilation should be free and unimpeded, assisted by fans if necessary. About fifty percent shade is adequate for growth.

Pleione formosana Hayata

(from Formosa)

This is probably the most widely grown species of *Pleione* and it is hardy enough to survive outdoors in some areas of temperate Australia. Native of China and Taiwan, it is variable in colour from lilac-purple and mauve through to white. This species has become a favourite because of its ease of culture and prolific flowering. A number of superior clones have been given cultivar names. Plants commonly cultivated as *P. pricei* are identical with this species.

Temp	Shade	In/on	Humidity	Air Flow
Cool	30-50%	Pot	50-70%	5

Pleione formosana

Pleione maculata (Lindley) Lindley

(spotted)

A native of northern India, Burma and Thailand, this species grows at lower altitudes than most other members of the genus, usually between 600 m and 2600 m. Because of this it requires more winter protection than other species. An autumn-flowering species the prominent white flowers are strongly marked with reddish-purple and have a bright yellow patch on the labellum.

Temp	Shade	In/on	Humidity	Air Flow
Cool-Int	30-50%	Pot	50-70%	5

Pleione maculata

Pleione praecox (Smith) D. Don

(flowering early)

Often grown as its synonym, *Pleione wallichiana*, this attractive species from northern India, Burma, southern China and Thailand, grows at elevations between 1500 m and 2700 m. It was one of the first species of the genus to be named but was originally classified as an *Epidendrum*. Being extremely variable in flower colour (white, pink, mauve, magenta, purple and even yellow), this species offers tremendous scope for selection. Plants flower in the autumn and the flowers have an attractive spicy perfume. The leaves of this species may take on colourful autumn tonings before falling.

Temp	Shade	In/on	Humidity	Air Flow
Cool	30-50%	Pot	50-70%	5

Pleione praecox

PLEUROTHALLIS R. Br.

(from the Greek *pleuron* a rib; *thallos* a branch; in reference to the tufting many-branched habit of these plants)

This generic name has become familiar to growers because of an upsurge in interest in their cultivation over recent decades. The term Pleurothallid has also become entrenched in the miniature orchidists' glossary, referring as it does to an alliance of closely related genera all of which lack pseudobulbs and have a single leaf atop a relatively slender stem. All are natives of the new world and the significance of the alliance is realised when it is considered that it numbers more than 1500 species. Until recently the genus *Pleurothallis* was considered to embrace a number of distinctive groups which are now regarded as separate genera. *Pleurothallis* itself consists of about 1000 species and they are found in Southern Florida, Mexico, the West Indies, Central America and South America. Most species grow as epiphytes on trees and rocks but a few species are terrestrials. While they are generally considered to be miniatures, it is surprising to learn that some species have unusual growths which may be more than one metre in length. The flowers of most species are small but of an intriguing shape and may be colourful or are variously spotted and blotched.

Cultivation: Species of *Pleurothallis* have become very popular with orchidists and are generally easy to grow. Temperature requirements depend on their origin, those from lowland areas needing warmth whereas the ones from mountainous regions will revel in cool conditions. Because they lack pseudobulbs, *Pleurothallis* plants should not be allowed to dry out; watering is daily in summer and two or three times a week in winter. Humidity should be continually high together with gentle air movement. Light requirements are generally not high but an enhancement of flowering is often noticeable in bright light. Mostly these orchids are grown in small pots but established plants can be successfully grown on slabs of treefern or oak branches. Fine particles of bark, charcoal, fern fibre and sphagnum moss are suitable for potting. Some growers use live sphagnum moss successfully on its own. These orchids have very fine roots which rot quickly if drainage and aeration becomes restricted and for the same reason they should not be overpotted. Repotting should take place every one or two years. Plants often decline fairly rapidly if conditions change and are not to their liking. First symptoms are leaf shedding or the excess production of plantlets. As many species can be easily propagated from leaf cuttings, it is a good procedure to propagate a few extra plants regularly.

Pleurothallis apthosa Lindley

(derivation unknown)

When viewed from the side, the unusual flowers of this orchid resemble a miniature clothes peg. With its petals and labellum minute, the whole flower is dominated by the sepals, the margins of which are minutely fringed. A native of Brazil, this species grows in warm, humid, mountainous forests.

Temp	Shade	In/on	Humidity	Air Flow
Int	30-50%	Small Pot	50-70%	5

Pleurothallis apthosa

Pleurothallis ghiesbreghtiana A. Rich. & Gal.

(after M. Ghiesbrecht, 19th century Belgian scientist)

The slender inflorescences of this orchid may grow to more than 30 cm long and they bear nodding, translucent flowers which are long lasting and have a sweet perfume. A native of the West Indies, Mexico and Central America, this species occurs in both sparse and dense forests up to 1500 m altitude. A variant is known in which the flowers self pollinate without opening.

Temp	Shade	In/on	Humidity	Air Flow
Int	30-50%	Small Pot	50-70%	5

Pleurothallis grobyi

Pleurothallis grobyi Bateman ex Lindley

(after Lord Grey of Groby, an English patron of horticulture)

With its neat clump of broadly ovate leaves and erect, slender racemes of cream to pale yellow flowers, this species makes an attractive plant for a small pot. Well grown specimens flower freely and produce an ornamental display. Of natural occurrence in Mexico and Central America, this species grows on trees in moist forests from sea level to about 900 m altitude.

Temp	Shade	In/on	Humidity	Air Flow
Int-Hot	30-50%	Small Pot	50-70%	5

Pleurothallis tribuloides (Sw.) Lindley

(like the genus *Tribulus*)

This species is one of the few in the genus to have bristly seed capsules. It can also be recognised by its brick red flowers borne one to three on short racemes which are mostly covered by prominent, whitish papery bracts. A rewarding species in cultivation, it is native to the West Indies, Mexico and Central America and it grows on trees in damp forests up to 1300 m altitude.

Temp	Shade	In/on	Humidity	Air Flow
Int	30-50%	Small Pot	50-70%	5

Pleurothallis ghiesbreghtiana

Pleurothallis tribuloides

POLYSTACHYA Hook.

(from the Greek, *poly*, many; *stachys*, a spike; in reference to the lateral branches found on the inflorescences of many species)

A very widespread genus which contains species which are distributed in Africa, Florida, Mexico, Central America, South America, India, China, South-east Asia and Indonesia. Totalling about 210 species, the genus includes a number of interesting and colourful orchids. In this country species of *Polystachya* are mainly regarded as collector's items, but many attractive species are deserving of wider recognition. They grow as epiphytes on trees or rocks in a wide variety of habitats including dense humid rainforests and open savannah where the climate is strongly seasonal.

Cultivation: It is difficult to generalise on the cultivation of such a diverse genus. Most of those species tried seem to adapt well to cultivation. Some species grow well on slabs, others in pots. Pots should be on the small side as overpotting is detrimental to the fine roots of these orchids. They are also very sensitive to clogging of the potting mix and poor drainage. Some species like shade, others bright light with all needing free air movement. Many growers favour fern fibres as potting materials for these orchids.

Polystachya adansoniae H. G. Reichb.

(growing on Baobab trees, *Adansonia*)

A native of tropical parts of Africa including Uganda, Kenya, Guinea and Angola, this species grows in fairly sparse vegetation often on Baobab trees. Usually sparse, plants of this species bear interesting, dense, almost cylindrical racemes in which the greenish flowers are borne upside down. Although mainly a collectors item, this species is easily grown in mixed collections.

Temp	Shade	In/on	Humidity	Air Flow
Int	10-30%	Slab	30-50%	5

Polystachya adansoniae

Polystachya zambesiaca

Polystachya zambesiaca Rolfe

(from the Zambesi River area)

The white labellum of this orchid provides a strong contrast to the widely spreading, yellowish-green sepals which in some variants are heavily suffused with brown. They are also finely hairy on the outer surface. The racemes arise on partially developed young growths and these fatten into pseudobulbs and produce new roots only when flowering is finished. A native of many African countries surrounding the Zambesi River and its tributaries, this species grows on trees and rocks. Plants have a definite dormant period after new growths have matured and at this stage should be kept on the dry side.

Temp	Shade	In/on	Humidity	Air Flow
Cool-Int	10-30%	Pot or Slab	30-50%	5

PROMENAEA Lindley

(after Promeneia, ancient Greek Preistess at Dodona)

A small genus of about 12 species all of which are native to Brazil. Originally most were included in the genus *Zygopetalum* and many have also been placed in *Maxillaria*. All are epiphytes with small, neat pseudobulbs and disproportionately large, colourful flowers borne singly on short racemes. Plants occupy very little space in an orchid collection and often flower quite freely.

Cultivation: Species of *Promenaea* are generally easy to grow and perform well in small pots of fern fibre. With their fine roots, plants are very sensitive to root rot and should be repotted immediately the drainage of the mixture becomes impaired. Warm conditions with high humidity and ample air movement are recommended. Bright, diffuse light promotes strong growth and flowering. A short rest period is beneficial after flowering has finished.

Promenaea stapelioides (Lindley) Lindley

(flowers resemble the genus *Stapelia*)

The lurid colouration in the flowers of this little Brazilian orchid always attracts attention, especially the velvety, purple-black cushion on the labellum and the bold striping on the sepals. With these unusual flowers and the dwarf habit which occupies little space, this species has become fairly popular with orchid collectors. Plants are easy to grow and flower freely.

Temp	Shade	In/on	Humidity	Air Flow
Int	30-50%	Small Pot	50-70%	5

Promenaea stapelioides

Promenaea xanthina (Lindley) Lindley

(golden yellow)

This species is still commonly known by orchid growers by the wrong name of *P. citrina*. A very popular orchid which is valued for its ease of culture, its modest neat habit and the relatively large, colourful flowers which are produced freely on well-grown plants. A native of Brazil it grows on protected trees and rocks in mountainous forests.

Temp	Shade	In/on	Humidity	Air Flow
Int	30-50%	Small Pot	50-70%	5

Promenaea xanthina

RESTREPIA Kunth

(after Don Jose E. Restrepo, 19th century South American naturalist)

The miniature orchids of this genus number about 30 species although many more names have been published because some species are highly variable. Natives of Mexico, Central America and South America, they reach their best development high in the Andes of Colombia and Ecuador. Conditions here are moist and humid throughout the year, one result of which is that these orchids have no need for storage organs and slender stems replace plump pseudobulbs. All species have a compact growth habit and the flowers, although usually small (some are about 5 cm long), are weird, interesting and often colourful. Some species hide their flowers on the undersides of leaves. For many years these orchids were submerged in the related genus *Pleurothallis* but they are a distinctive group in which the lateral sepals are fused to form a platform and the dorsal sepals have filamentous extensions which end in fleshy knobs or clubs.

Cultivation: These orchids, which are easy to grow, do best in small pots and as a general rule should not be overpotted. Some growers mount established plants on tree fern slabs or pieces of oak. They grow well under a range of conditions and may flower several times a year. If happy, plants will increase in size at an astounding rate. Potting materials include fern fibres, fine particles of bark, charcoal and chopped sphagnum moss. Small rootless pieces can be started well in live sphagnum moss. These orchids do not like to dry out and should be watered daily during summer and two or three times a week in winter. Humidity should be high with good air movement. When plants begin to lose leaves or produce masses of plantlets instead of new growths it is a sure sign they are in a decline. As they are readily propagated from leaf cuttings and plantlets, it is always a good idea to keep a few propagations underway.

Restrepia elegans

Restrepia elegans Karsten

(elegant)

This exquisite Venezuelan endemic is one of the more commonly grown species of this unusual genus. Its stiff, broadly elliptical, light green leaves are borne singly atop slender stems clothed with overlapping white papery sheaths. Plants form a neat, dense tuft and the delightfully spotted flowers are borne on long, filiform stems which arise from a sheath at the leaf base. Each flower is about 5 cm long.

Temp	Shade	In/on	Humidity	Air Flow
Int	30-50%	Pot	50-70%	5

Restrepia hemsleyana Schltr.

(after William Botting Hemsley, former keeper of Kew Herbarium)

A miniature orchid with relatively large flowers, this species is native to Venezuela and Colombia, growing in mountainous rainforest at about 2000 m elevation. With the plants forming tufts of slender stems each topped by a relatively thick, keeled dark green leaf about 7 cm long, they can be readily accomodated in a small pot. Dark red flowers, each about 5 cm long, are borne on long slender stems which often appear to arise from beneath the leaf. Plants flower at intervals throughout the year.

Temp	Shade	In/on	Humidity	Air Flow
Int	30-50%	Pot	70-90%	5

Restrepia hemsleyana

RHYNCHOLAELIA Schltr.

(literally a *Laelia* with a beak)

A small genus of 2 or 3 species which are native to Mexico and Central America. They grow as epiphytes or terrestrials in relatively sparse forests and are valued by collectors for their ease of culture and showy flowers. At various times both species have been included in the genera *Laelia* and *Brassavola*.
Cultivation: These orchids are very easy to grow and thrive in conditions similar to those required by Cattleyas, that is, warmth, relatively high humidity and regular gentle air movement. Best flowering is achieved in bright, diffuse light. Plants are not fussy as to potting materials providing that drainage is free and unimpeded.

Rhyncholaelia digbyana (Lindley) Schltr.

(after St Vincent Digby, 19th century English orchid grower in whose collection the species first flowered)

Renowned for the remarkable deep fringing of its labellum, this lovely orchid is deservedly popular in cultivation. Although the flowers appear very delicate, in reality they are quite tough and long lasting. They are also strongly fragrant at night. This species occurs naturally in Mexico, Honduras and Guatemala. A variant, with delicate fringing on the petals, (var. *fimbripetala* Ames) is known from Honduras.

Temp	Shade	In/on	Humidity	Air Flow
Int-Hot	30-50%	Pot or Slab	50-70%	5

Rhyncholaelia digbyana

Rhyncholaelia glauca (Lindley) Schltr.

(bluish-grey)

A very popular orchid prized by growers for its ease of culture and free-flowering habit. Of natural occurrence in Mexico, Guatemala and Honduras, it grows on trees (occasionally a terrestrial) in sparse mountain forests up to 1500 m altitude. The large, heavy-textured, long-lived flowers are delightfully fragrant and well-grown plants may have flowers present over many months of the year.

Temp	Shade	In/on	Humidity	Air Flow
Int	10-30%	Pot	30-50%	5

Rhyncholaelia glauca

ROSSIOGLOSSUM Garay & G. C. Kennedy

(a derivative of *Odontoglossum*)

A small genus of about 4 species of orchids all originating in Mexico and Central America. They are epiphytes with large, colourful flowers and were previously included in the genus *Odontoglossum*. Plants of all species are very similar with laterally flattened, somewhat sharp-edged pseudobulbs and relatively short leaves.

Cultivation: These orchids like cool to intermediate conditions, high humidity and an abundance of gentle air movement. They are also shade lovers and plants tend to burn badly if exposed to excessive light. Plants can be grown on slabs of treefern but mostly they are grown in pots of fern fibre or softwood bark, charcoal and sand. Drainage of the potting mix must be free and unimpeded. Plants have only a short quiescent period after the growths mature and should be kept regularly moist throughout the year.

Rossioglossum grande

Rossioglossum grande (Lindley) Garay & G. C. Kennedy

(large, big)

The large, colourful flowers of this lovely orchid always attract attention. A native of Mexico and Guatemala this species grows on trees in ravines and other shady locations in forests up to 2700 m altitude. Since its discovery in 1839 it has become one of the most popular orchids grown although many plants are killed needlessly when housed in unsuitable conditions or when the potting mix deteriorates. The flowers, which can be up to 16 cm across have a waxy lustre and are long lasting.

Temp	Shade	In/on	Humidity	Air Flow
Cool-Int	30-50%	Pot	50-70%	5

Rossioglossum schlieperianum (H. G. Reichb.) Garay & G. C. Kennedy

(after Herr Schlieper, 19th century German orchid grower in whose collection this species first flowered)

A native of Costa Rica, this species has flowers of a similar colouration to *R. grande* but they are smaller and more are carried on each raceme. The much narrower, less patterned labellum of *R. schlieperianum* is a useful diagnostic guide. This species adapts better to warm conditions than does *R. grande*.

Temp	Shade	In/on	Humidity	Air Flow
Int	30-50%	Pot, Basket or Slab	50-70%	5

Rossioglossum schlieperianum

Sophronitis mantiqueirae

SOPHRONITIS Lindley

(from the Greek *sophron*, modest; in reference to the neat growth habit of these orchids)

A small genus of 7 species, all endemic to Brazil. Prized for their neat growth habit and disproportionately large and highly colourful flowers, these orchids have captured the imagination of growers world wide. Although of delicate appearance, in reality these are tough little orchids which grow naturally in conditions of bright light (often full sun for at least part of the day) and usually near some source of high humidity (streams, ponds or colonies of terrestrial bromeliads) and with constant fresh air. Some species grow naturally on rocks, others prefer the trunks of small trees. Those species from the lowlands relish warm, humid conditions whereas the others, which grow at high altitudes, experience much cooler conditions usually with cold nights (to freezing point in winter) and these species languish if placed in warm conditions.

Cultivation: Even experienced orchid growers admit that the successful cultivation of *Sophronitis* species is tricky and unless plants are grown well, high quality flowering is unlikely to be achieved. However these little orchids reward successful growers wih dazzling displays of brilliantly coloured flowers and this is sufficient incentive for extra care in their culture. Plants of these orchids like bright light and in natural habitats often grow in full sun with the thicker than normal leaves taking on reddish hues. Humidity should be high with constant, bouyant air movement and while some species like warm to hot conditions, those originating in mountainous regions need cooler conditions and will linger and die unless these are provided. Various potting mixes are used by growers with considerable success being achieved in fresh sphagnum moss and slightly less success in fern fibre. Mounting on slabs of cork or English Oak can be successful. Liquid fertilisers are very beneficial to these orchids. *Sophronitis* species react quickly to good growing conditions so do not be afraid to move them around until their optimum microclimate is achieved.

Sophronitis mantiqueirae (Fowlie) Fowlie

(from the Serra da Mantiqueira Mountains, Brazil)

This species has been confused for many years with *S. coccinea* but plants have shorter pseudobulbs and its leaves are either wholly reddish or spotted with red on the underside (a prominent red line along the dorsal leaf vein is an easy means to distinguish plants of *S. coccinea*); also the labellum of the flower has three to five dark veins (seven to nine in *S. coccinea*). *S. mantiqueirae* occurs naturally on some Brazilian mountains between 1300 m and 2000 m altitude. Plants grow on small, moss-covered trees in situations of bright light and where air movement is abundant.

Temp	Shade	In/on	Humidity	Air Flow
Cool-Int	10-30%	Pot, Basket or Slab	50-70%	5

Sophronitis wittigiana Barb. Rodr.

(derivation unknown)

Plants of this species are easily recognised by their nearly globular pseudobulbs being arranged in two close ranks; each new growth being produced on the side alternate to its predecessor. The flowers also are of a lovely pink colouration not seen in any other species of *Sophronitis*. A native of Brazil, this highly ornamental species grows at elevations of about 1200 m on trees close to perched swamps.

Temp	Shade	In/on	Humidity	Air Flow
Int-Hot	10-30%	Pot or Slab	50-70%	5

Sophronitis wittigiana

SPATHOGLOTTIS Blume

(from the Greek spathe, a blade; glotta, a tongue; in reference to the prominent blade-like labellum)

A genus of terrestrial orchids with colourful flowers which have a very distinctive cruciform labellum. Consisting of about 50 species, the genus is widely distributed from India and China to Malaysia, Indonesia, New Caledonia, various Pacific islands, New Guinea and Australia. Most species grow in relatively open habitats, frequently in moist soils. Many are deciduous, shedding their leaves to avoid extremes of dryness, but some species from high altitudes appear to be evergreen. These orchids are widely cultivated in Asia where they are common garden plants prized for their colourful flowers. Colours include whites, pinks, mauves, purples and yellows.

Cultivation: In the tropics many of these orchids can be used as garden plants and they are prized for their ease of culture and long flowering period. In temperate regions they must be sheltered in the warmth and protection of a glasshouse. Here plants are grown in pots of a fibrous terrestrial mix containing some good quality loam and leaf mould. They need bright light and should be watered copiously when in growth but sparingly as they become dormant and shed their leaves.

Spathoglottis pubescens

Spathoglottis pubescens Lindley

(slightly hairy)
Buttercup Orchid

Widely distributed in India, China, Burma and Hong Kong, this attractive terrestrial orchid is commonly found growing on hills and slopes from sea level to about 800 m elevation. It forms compact clumps often among grasss and in quite exposed situations. The climate is strongly seasonal and plants shed their leaves completely during the dry season. While the flowers are commonly bright yellow in some areas they are marked with orange or reddish streaks on the exterior.

Temp	Shade	In/on	Humidity	Air Flow
Int-Hot	10-30%	Pot	30-50%	4

STANHOPEA Frost ex Hook.

(after the Right Honorable Philip Henry, 4th Earl of Stanhope)

The accurate identification of cultivated plants of this genus is fraught with difficulty since the species are quite variable in colour, floral markings and even the shape of some of the floral parts. This variation has led to more than 120 names being published for what seems to be a genus of about 30 species (although botanists cannot agree on the exact number). These problems of taxonomy must not be allowed to detract from what is an amazing genus of quite astonishing orchids. Their large, waxy, almost bizarre flowers never fail to arouse comment, whether it be caused by their sheer size, waxy texture, colouration, unusual shape or the powerful perfumes these orchids release.

All species have a similar growth habit of crowded conical pseudobulbs and large, leathery leaves. Native to Mexico, Central America and South America they commonly grow on trees or rocks, but incredibly a few species grow in the soil. Their racemes are stiffy pendant so that the flowers are projected downwards and thus the plants are only suited to basket culture or on slabs. Their buds develop with incredible speed and when sufficiently swollen and turgid they open with a distinct snapping or popping sound. Their flowers are short-lived usually lasting only three or four days, during which time they fill a glasshouse with an almost overpowering perfume.

Cultivation: Species of *Stanhopea* adapt very well to cultivation and are generally very easy orchids to grow. As already mentioned baskets or slabs are necessary for their culture. They require a well-drained compost of medium sized particles of such materials as pine bark, fern fibre, charcoal and coarse sand or perlite. Plants need bright light, warmth, humidity and air movement. Watering should be copious while in active growth but when the pseudobulbs fill out the plants should be kept on the dry side for four to six weeks to enhance flowering.

Stanhopea insignis Frost

(remarkable)

The type species of the genus, this handsome orchid was first collected in the 1820's and is native to Brazil and Peru. Its typically pendulous, dull yellow, densely spotted flowers are heavily fragrant and short-lived.

Temp	Shade	In/on	Humidity	Air Flow
Int	10-30%	Basket	30-50%	5

Stanhopea insignis

Stanhopea oculata (Lodd.) Lindley

(with an eye-like patch of colour)

In its native state, this orchid is sometimes recorded growing as a terrestrial; such plants must present an unusual appearance when in flower. More frequently it grows on trees or rocks in humid forests and occurs naturally in Mexico, Central America and South America. The short-lived flowers release a strong, vanilla-like fragrance.

Temp	Shade	In/on	Humidity	Air Flow
Int	10-30%	Basket	30-50%	5

Stanhopea tigrina

Stanhopea wardii Lodd. ex Lindley

(after M. Ward, original collector)

A commonly grown species which is native to Mexico and Central America where it grows on trees in humid forests. Plants are prized for their spectacular flowers each of which is about 12 cm across and they are borne up to nine at a time on a pendulous raceme about 30 cm long. The flowers open greenish-cream and quickly age to pale yellow. They are plentifully spotted with reddish-purple and there are two large prominent purplish-brown spots on the labellum. The flowers diffuse a heavy fragrance into the atmosphere and this quickly fills a glasshouse to such an extent as to be almost overpowering. Although spectacular, the flowers usually only last three or four days. This species grows very easily in a basket of coarse, epiphytic mixture.

Temp	Shade	In/on	Humidity	Air Flow
Int	30-50%	Basket or Slab	50-70%	4

Stanhopea oculata

Stanhopea tigrina Bateman ex Lindley

(marked like a tiger)

A very familiar species which is native to Mexico and may also occur elsewhere but its exact range seems to be uncertain due to confusion with other species, in particular *S. hernandezii* and *S. nigroviolacea*. It has larger flowers than either of these species and the dark purplish black colouration of the petals in particular is most distinctive.

Temp	Shade	In/on	Humidity	Air Flow
Cool-Int	10-30%	Basket	20-30%	5

Stanhopea wardii

STELIS Sw.

(from an ancient Greek word used by Theophrastus for a parasitic plant like mistletoe)

A group of intriguing miniature orchids which numbers about 500 species. With a major concentration in the Andes of South America, these orchids also occur in Central America, Mexico and the West Indies. They grow as epiphytes on mossy trees and rocks. Most species are fairly small with a compact growth habit and tiny triangular, flowers usually carried in elongated racemes. In some the flowers are light sensitive, opening only in bright light or sunshine and closing at night. They lack pseudobulbs since the conditions where they grow are moist and humid for most of the year.

Cultivation: Generally easy to grow, these orchids have similar requirements to the closely related genera *Pleurothallis* and *Restrepia*. They are best grown in small pots although established plants can be attached to slabs of treefern or oak branches. Fine particles of bark, charcoal, fern fibre and sphagnum moss are suitable for potting. Species originating in the lowlands require warmth whereas high altitude species are cool growing. All species need regular watering, high humidity and gentle air movement.

Stelis argentata Lindley

(silvered)

When viewed under the lens, the tiny flowers of this species seem to sparkle like a jewel. Borne on slender racemes which arise from the base of the leaf, the flowers can never be called showy but are always of interest. The prominent sepals, which make up the bulk of the flower, have a fine fringe of marginal hairs. Flower colour ranges from silvery green to reddish. The species is native to Mexico and various countries in Central America and South America.

Temp	Shade	In/on	Humidity	Air Flow
Int	30-50%	Small Pot	50-70%	5

Stelis argentata
Stelis argentata (flower)

Stenoglottis fimbriata

STENOGLOTTIS Lindley

(from the Greek, *stenos*, narrow; *glottis*, mouth of the windpipe; a reference to the narrow division of the labellum)

A genus of 3 or 4 species all native to Africa where they are mainly found in southern areas. Commonly of terrestrial habit, plants of these orchids grow in shady forests and on earthern embankments, but they are also often grow on rock outcrops and one species sometimes grows as an epiphyte on mossy tree trunks. Usually clump-forming, these robust orchids have decorative leaves (often spotted or blotched) and colourful flowers crowded in tall racemes.

Cultivation: Members of this small genus generally resent repotting and are slow to recover from such an occurrence. Consequently they are best grown as specimen plants and they do not seem to mind being crowded together in clumps. The plants have large, fleshy, woolly roots which can completely fill a pot before repotting becomes absolutely essential. Crowded clumps of these orchids flower more freely than single plants and a potful can produce an impressive display. Well-drained terrestrial mixes or epiphytic mixes composed of relatively small particles are suitable for their culture. After flowering, plants have a definite dormant period during which they should be kept on the dry side until new shoots appear. These orchids are hardy enough to be grown as garden plants but they must be continually protected against slugs and snails which relish their succulent tissues.

Stenoglottis fimbriata Lindley

(fringed)

A native of eastern South Africa (Natal, Transvaal, Zimbabwe, Malawi and Tanzania), this species commonly grows as a terrestrial, often near rocks, but may be sometimes found as an epiphyte on moss-covered tree trunks. Decorative features of this species include bright green leaves attractively flecked with brownish-purple markings and tall spikes crowded with lilac to mauve, spotted flowers which have an interesting, deeply lobed labellum.

Temp	Shade	In/on	Humidity	Air Flow
Cool	30-50%	Pot	30-50%	3

Stenoglottis longifolia J. D. Hook.

(with long leaves)

Although similar in general appearance to *S. fimbriata*, this species is generally more robust with up to eighty flowers in a taller raceme and the labellum has five prominent lobes (three in *S. fimbriata*). An excellent pot plant which produces impressive floral displays, this species grows naturally on rocks in Natal, southern Africa.

Temp	Shade	In/on	Humidity	Air Flow
Cool	10-30%	Pot	30-50%	4

Stenoglottis longifolia

ZOOTROPHION Luer

(from the Greek *zootrophion*, a menagerie; alluding to the similarity of the flowers to the heads of various animals)
Window Orchids

The common name of 'Window Orchids' is apt for members of this genus because the flowers have a gap in the side of the united dorsal sepal and lateral sepals through which the internal parts can be seen and where the pollinating agent enters. Miniature orchids closely related to *Pleurothallis*, they number about 11 species and are of natural occurrence in countries of Central America and South America where they grow on trees or rocks in humid, shady conditions.

Cultivation: These orchids have similar cultural requirements to species of *Pleurothallis* and *Restrepia*. They are best grown in small pots of fibrous mix or sphagnum moss. Drainage must be excellent. Plants have no storage capacity and must never be allowed to become severely dry. Warm, humid conditions with regular, gentle, unimpeded air movement are ideal.

Zootrophion atropurpureum (Lindley) Luer

(dark purple, in reference to the flowers)

Originally discovered in 1836 in Jamaica, this species created considerable interest among English naturalists, botanists and orchid hobbyists when introduced in 1843. Taxonomically this species has had a chequered history, first being described in the genus *Specklinia*, then transferred to *Masdevallia*, *Pleurothallis* and *Cryptophoranthus* before being placed in its present genus. The unusual flowers aroused the interest of Charles Darwin who carried out detailed studies to elucidate the pollination mechanism and reported his results in his fascinating book *The Various Contrivances by which Orchids are Fertilised by Insects*, John Murray, London 1882. *Z. atropurpureum* is fairly easily grown in a small pot of fine mix or sphagnum moss.

Temp	Shade	In/on	Humidity	Air Flow
Hot	50-70%	Pot or Slab	70-90%	4

Zootrophion atropurpureum

ZYGOPETALUM Hook.

(from the Greek *zygos*, a yoke; *petalon*, a petal; the swelling at the base of the labellum appears as a yoke)

A genus of about 22 two species most of which occur naturally in Brazil, but with some also known from Paraguay, Peru, Bolivia and Venezuela. At various times other genera such as *Pescatorea*, *Huntleya*, *Chondrorhyncha* and *Promenaea* have been included in *Zygopetalum* but are now regarded as being distinct. True species of *Zygopetalum* have prominent pseudobulbs and thick-textured flowers borne on multiflowered racemes. Plants grow as terrestrials or epiphytes on rocks and trees and have truly beautiful flowers which always capture attention. Despite the beauty of their flowers and ease of culture, few species have become well entrenched in Australian orchid collections.

Cultivation: Few orchids are easier to grow than the common species of *Zygopetalum* and these make excellent subjects for beginners. Most of the less-common species seem to be adaptable although some, such as *Z. maxillare*, may have specialised requirements. Most species are strong growers with coarse roots and they thrive in a variety of potting mixes and respond well to the application of fertilisers. Although growth is satisfactory in shady conditions, flowering is enhanced by bright, diffuse light, some species even tolerating short periods of sunshine. Gentle, free, air movement is very important for healthy plants.

Zygopetalum crinitum Lodd.

(long-haired like a mane)

Easily confused at first glance with *Z. mackaii*, plants of this species are less robust, the flower spikes often spread horizontally and the veins on the labellum are fringed with prominent bristles. Of natural occurrence in Brazil, it grows as an epiphyte on the main branches of rainforest trees. The colourful, long-lived flowers have an attractive perfume. The prominent veins on the labellum are variable in colour from red to pink or blue.

Temp	Shade	In/on	Humidity	Air Flow
Cool-Int	30-50%	Pot	30-50%	5

Zygopetalum intermedium Lodd.

(intermediate between others)

This, the most commonly cultivated member of the genus, is usually wrongly grown as *Z. mackaii*, the latter being very rare in Australian collections. *Z. intermedium* can be distinguished by its column front and base of labellum being hairy and the petals being of similar length to the dorsal sepal. A true beginners orchid, this species is adaptable, easy to grow and free flowering with the heavy-textured flowers being long lasting and delightfully fragrant. They also last well after cutting and in the USA, plants are grown commercially for their cut flowers. In warm temperate and subtropical areas potted plants can be grown outdoors under the scant protection of shrubs or trees.

Temp	Shade	In/on	Humidity	Air Flow
Cool-Int	10-30%	Pot	30-50%	5

Zygopetalum crinitum

Zygopetalum maxillare Lodd.

(shaped like a jawbone)

In its native state this species grows exclusively on the trunks of tree ferns in forested gullies and on humid slopes. It occurs naturally in Brazil and Paraguay and has become sought after by orchid growers because of its rich-coloured labellum and showy floral displays. Plants however can be difficult to maintain in cultivation with some success being achieved on slabs of treefern or in pots of fern fibre. Drainage must be excellent.

Temp	Shade	In/on	Humidity	Air Flow
Int	30-50%	Pot	50-70%	5

Zygopetalum triste Barb. Rodr.

(dull-coloured)

As the photograph shows clearly, the specific name for this orchid is obviously ill-chosen for its flowers are very colourful indeed. Plants have relatively small pseudobulbs and the flowers tend to be crowded towards the end of the racemes. A native of Brazil it grows on wet rocks at rather high altitudes. In Australia, this species is mainly a collector's item.

Temp	Shade	In/on	Humidity	Air Flow
Int	30-50%	Pot	50-70%	5

Zygopetalum intermedium

Zygopetalum maxillare

Zygopetalum triste

161

Glossary

abberant Unusual or atypical; differing from the normal form

acuminate Tapering into a long, drawn-out point

acute Bearing a short, sharp point

adnata Attached by the whole length or a substantial part

adventitious Arising in an unusual position, often said of adventitious roots or buds

aerial roots Adventitious roots arising on stems and growing in the air; such roots are often prominent on monopodial orchids.

aff., affinity A botanical reference used to denote an undescribed species closely related to an already described species

alternate Borne at different levels in a straight line or in a spiral

antennae Slender elongated appendages on flowers; a term often used for the erect petals in the flowers of *Dendrobium*, section Spatulata; sometimes used for the filiform appendages on the labella of *Phalaenopsis*

anther The pollen-bearing part of a stamen

anther cap The cap-like structure which terminates the column and covers the pollinia

anthesis The period of flowering

apical At the apex

apiculate Ending in a short, sharp point

apomixis The production of seeds without the union of sex cells; this is a process of vegetative reproduction

articulate Jointed; with an abscission layer

asexual propagation Propagation by vegetative means; for example, division, aerial growths, meristem culture

attenuate Drawn out, tapering

autogamy The process of self-pollination

axil Angle formed between adjacent organs in contact; commonly applied to the angle between a leaf and the stem

axillary Borne within the axil

axis The main stem of a plant or part of a plant

backbulb The older pseudobulbs which are usually leafless; pseudobulbs often have viable buds and may produce a new shoot when severed from the plant.

basal Arising from the base; often said of the point where an inflorescence arises

beak A point; often used for the point on the anther (rostrum)

bifid Deeply notched for more than half its length

bifurcate Forked or notched

bigeneric A term used for hybrids between two genera

bilobed Two-lobed

bisexual Both male and female sexes present

blade The expanded part of a leaf or labellum

blotch A large spot of colour of irregular shape

botanical A term used by orchid growers for small-flowered species of limited interest

bottom heat A propagation term used to denote the application of artificial heat in the basal region of the division

bract A leaf-like structure which lacks a blade or lamina.

bristly With stiff hairs or bristles

bud An unopened flower or a new shoot in its early stages

calcareous An excess of lime, as in soil

calli Non-secreting glands found on orchid labella; in many terrestrial orchids they are associated with deceit or mimicry

callus A fleshy or plate-like structure found on the labellum; it may have ridges or other outgrowths or be associated with nectar production

calyx All of the sepals of a flower

cap A familiar term for the anther cap

capsule A dehiscent, dry fruit containing many seeds

carpel Female reproductive organ

caudate With long tail-like appendages or filiform tips; as in the floral segments of *Bulbophyllum masdevalliaceum*

cauline Belonging to a stem, usually referring to leaves

chlorophyll The green pigment of leaves and other organs, important as a light-absorbing agent in photosynthesis

chlorotic A yellowish plant deficient in nitrogen or iron

ciliate With a fringe of fine hairs

cirrhus A term sometimes used for the filiform appendages on the labella of *Phalaenopsis* flowers

clavate Club-shaped; thickened towards the apex

claw Another term for a narrow stalk on the base of a segment

cleistogamy The process of self-pollination occurring without the flowers opening

clinandrium The apical margins of the column or the cavity where the anther fits

clone A group of plants propagated vegetatively from one plant (usually a superior form); all members of a clone are genetically identical

club A familiar term used for the expanded apical part of sepals or petals

clubbed A term used for thickened segments; see clavate

column the central, fleshy structure in orchid flowers composed of the style and staminal filaments

column foot An extension (usually fleshy) of the base of the column

column wing A flattened, often wing-like appendage of the column which is probably a sterile stamen

compact Short, reduced in length; some epiphytic orchids have forms with a compact growth habit

compressed Flattened laterally

concave Sunken, basin-like, often used to describe a stigma

conduplicate Folded together along its length, with each half flat

congeneric Belonging to the same genus

congested Crowded closely together

conical Cone-shaped

constricted Narrowed or drawn together at some point

contorted Twisted

contracted Narrowed

convex Curving outwards

convolute Rolled up lengthwise; referring to the way some leaves are folded when young

cordate Heart shaped

coriaceous Leathery in texture

corm A thick underground stem, of several internodes; as in the pseudobulbs of *Geodorum neocaledonicum*

crenate The margin cut regularly into rounded teeth

crest A term used for the callus of some orchids

crisped The margins very wavy or crumpled, as in the labella margins of many orchids

cross-fertilisation Fertilisation by pollen from another flower

cross-pollination Transfer of pollen from flower to flower

cultivar A horticultural variety of a plant or crop

cupped When the segments remain concave and do not become flat

cylindrical Round in cross-section and not tapered lengthwise

cyme An inflorescence where the branches are opposite

cytoplasm The living material within a cell

damping-off A condition in which young seedlings are attacked and killed by soil-borne fungi

deciduous Falling or shedding of any plant part

dehiscent Splitting or opening when mature

dentate Toothed

depauperate A weak plant or one imperfectly developed

determinate Said of a growth or inflorescence when it has an extension limit

diandrous With two anthers

dichotomous Forking regularly into two equal branches or parts

dicotyledon Any Angiosperm which has two seed leaves and reticulate venation in the leaves

diffuse Widely spreading and much branched; of open growth

dimorphic Existing in two different forms

diploid With two sets of chromosomes

disc The central portion of the labellum where the lobes meet; often the area where the callus is developed

dissected Deeply divided into segments

distal Away from the base towards the apex

distichous In two ranks, usually applied to the arrangement of leaves or flowers

diurnal During the day, as describing flowers that only open in the day

dormancy A physical or physiological condition that prevents growth or germination even though external factors are favourable

dorsal The upper side; complicated in orchid flowers because of resupination; the dorsal sepal is the odd one of the three sepals but it is not always on the upper side of the flower

downy Covered with soft hairs

duplicate Leaves which are folded once along the centre, the two halves being flat

ecology The study of the interaction of plants and animals within their natural environment

ellipsoid Spindle-shaped, tapering to each end in three dimensions

elliptic Oval and flat in a plane, narrowed to each end which is rounded

entire Whole; not toothed or divided in any way

emarginate With a shallow notch at the apex

endemic Restricted to a particular country, region or area

endosperm Tissue rich in nutrients which surrounds the embryo in seeds; orchid seeds lack endosperm

ensiform Sword-shaped, as in the leaves of *Cymbidium ensifolium*

ephemeral Short-lived; in flowers referring to those which last a few hours or less

epiphyte A plant growing on or attached to another plant but not drawing nourishment from it and therefore not parasitic

equitant Said of laterally flattened leaves arranged in two ranks that overlap at the base

erect Upright

erose With an irregularly cut or notched margin as if chewed

evergreen Remaining green and retaining leaves throughout the year

exotic A plant introduced from overseas

eye A term used for a viable vegetative bud

falcate Sickle shaped

family A taxonomic group of related genera

fertile bract A bract which subtends a pedicel

fertilisation The act of union of the male gametes (from the pollen) with the egg cells in the ovules

filament The stalk of the stamen supporting the anther

fimbriate Fringed, especially along a margin

floriferous Free-flowering

foetid A disagreeable odor

foliaceous Leaf-like

forked Divided into two equal segments

free Not joined to any other part

fruit The seed-bearing organ developed after fertilisation

furrowed Grooved longitudinally

fusiform Spindle shaped; widest in the middle and tapered to each end

genus A taxonomic group of closely related species

geophyte A plant growing in the ground

germination The active growth of an embryo resulting in the development of a young plant

glabrous Without hairs

gland An organ which secretes fluid

glandular Bearing glands

glaucous Covered with a bloom giving a bluish lustre

globoid Globe-like, globular, spherical

globose Globular; almost spherical

glutinous Very sticky

gynandrium Another term for the column

gynoecium The female parts of a flower

gynostegium Another term for the column

habit The general appearance of a plant

habitat The environment in which a plant grows

haploid With one set of chromosomes

head An inflorescence with the flowers in a tight cluster; for example, *Tropidia curculigoides*

herb A plant which produces a fleshy rather than a woody stem

herbaceous A perennial plant which dies down each year after flowering

herbarium A botanical collection of pressed plant specimens

hermaphrodite Having male and female parts on a flower

hirsute Covered with long, spreading coarse hairs

humus The friable layer on the soil surface formed from decaying vegetation

hyaline Translucent or transparent

hybrid The progeny of a cross between two species, cultivars or other hybrids

hybridisation The act of crossing flowers to produce hybrids

imbricate Overlapping like fish scales

incised Deeply and irregularly cut

indehiscent Not splitting open at maturity

indeterminate Said of a growth or inflorescence when it has no apparent extension limit

indigenous Native to a country, region or area

inflorescence The flowering structure of a plant

internode The part of a stem between two nodes

introrse Turned inwards towards the axis

jointed Bearing distinct joints or nodes

keel A ridge like the base of a boat; such ridges may be a common adornment on the labellum callus of orchids; the midrib on the underside of the leaves of many orchids protrudes as a keel

labellum A lip; in orchids and gingers the petal in front of the flower

lacerate Appearing as if irregularly cut or torn

lamina The expanded part of a leaf or labellum

lanceolate Lance-shaped; longer than wide and tapering at each end, especially the apex

lateral Arising from the main axis; arising at the side of the main axis

lateral lobes The two side lobes of a labellum

lax Loose, drooping, non-turgid

lead A term used by growers for a new growth

leaf-base Specialised and expanded basal part of the petiole where it sheathes the pseudobulb or stem

leafless Lacking leaves

ligulate Strap-shaped

linear Long and narrow with parallel sides

lingulate Tongue-shaped

lip See labellum

lithophyte A plant that grows on rocks, boulders and cliff faces

littoral Growing in communities near the sea

lobe A segment of an organ as the result of a division

lorate Strap-shaped

marginal Attached to or near the edge

mealy Covered with flour-like powder

medium The potting mix in which an orchid is grown, or the mixture on which seeds are raised

membranous Thin-textured

mentum A chin-like extension at the base of the flower consisting of the column-foot and the bases of the lateral sepals

meristematic Tissue which retains the capacity for further growth

mid-lobe The main projecting lobe of a labellum

midrib The principal vein that runs the full length of a leaf or segment

mimicry A deceitful resemblance between different organisms

monandrous With one anther

monocotyledon Any Angiosperm which has one seed leaf and parallel venation in the leaves

monopodial A stem with a single main axis which grows forward at the tip

monotypic A genus with a single species

mucronate With a short, sharp apex (or mucro) on a leaf

mycelium A mass of fungal strands

mycorrhiza A beneficial relationship between the roots of a vascular plant and fungi resulting in nutrient exchange

mycotropic Another term for saprophytic

naked A term used for pollinia which lack any supporting structures such as stipes or caudicles

nectar A sweet fluid secreted from a nectary

nectar guide Markings on petals or the labellum which lead a pollinator to the nectar

nectary A gland which secretes nectar

nerves The fine veins which traverse the leaf-blade

node A point on the stem where leaves or bracts arise

obcordate Cordate with the broadest part above the middle

oblanceolate Lanceolate with the broadest part above the middle

oblong Longer than broad, with parallel sides and rounded ends

obovate Ovate with the broadest part above the middle

obtuse Blunt or rounded at the apex

offset A growth arising from the base of a plant and producing roots while still attached.

opposite Arising on opposite sides but at the same level

orbicular Circular in outline

orchidist A person very interested in orchids and their cultivation

orchidologist A person who studies orchids

orchidology The study of orchids

ovary The part of the gynoecium which encloses the ovules and after fertilisation develops into the fruit

ovate Egg-shaped in a plane

ovoid Egg-shaped in three dimensions

ovule The small structure within the ovary which becomes a seed after fertilisation

panicle A much-branched racemose inflorescence

paniculate Arranged in a panicle

parasite A plant that derives nourishment directly from another living plant

pedicel The stem which supports a single flower in an inflorescence

peduncle The main axis of a compound inflorescence or the stalk of a solitary flower which subtends the pedicel

peloric An abnormality whereby the labellum is of a similar shape and colour to the other petals

pendent Hanging downwards

perennial A plant living for more than two years

perianth A collective term for the petals and sepals of a flower

petal A segment of the inner perianth whorl or corolla

petaloid With the appearance of a petal; reversions of this type are common in some orchids

petiole The stalk of a leaf

pilose Bearing long, soft hairs

pistil A collective term for the female organs of a flower (stigma, style and ovary)

pleated Folded longitudinally

plicate Folded longitudinally in pleats

pollen The one-celled male spores that are borne in the anther

pollinarium The whole male structure as moved by an insect during pollination

pollination The transference of pollen from the anther to the stigma of a flower

pollinium (pl. pollinia) An aggregated, coherent mass of pollen grains found in the Orchidaceae; usually the contents of an anther cell

polymorphic Consisting of many forms; a variable species

proliferous Bearing offshoots and other processes of vegetative propagation

protocorm A roughly spherical body that develops after germination and from which a shoot develops after mycorrhizal infection

protoembryo A term used for orchid embryos because they lack any differentiation into tissues

pseudobulb Thickened bulb-like stems of sympodial orchids bearing nodes

pseudobulbous Having pseudobulbs

pseudocopulation A type of mimicry whereby flowers deceive male wasps into attempting copulation in order to achieve pollination

pubescent Softly hairy

pyriform Pear shaped

raceme A simple, unbranched inflorescence with stalked flowers

rachis See rhachis

recurved Bent backwards

reed-stem Said of some orchids which have slender, leafy stems of uniform thickness, for example, *Dendrobium lobbii*

respiration The process by which living cells oxidise food compounds to produce energy, water and carbon dioxide

resupinate Said of an orchid flower which has the labellum on the lower side; in bud such flowers are upside down and they right themselves by twisting

reticulate With veins which interconnect like a net

retrorse Pointed strongly backwards towards the base

retuse The apex rounded and with a shallow notch

revolute With the margins rolled back

rhachis The main axis of a compound leaf or an inflorescence (to which the pedicels are attached)

rhizome An underground stem with nodes, roots and which can form shoots.

rosette When the basal leaves radiate roughly in a circle from a central axis

rostellum The area of tissue that separates the stigma from the anther; an adhesive portion of the stigma which aids pollen transfer

saccate Deeply pouched; like a sack

saprophyte A leafless (or nearly so) plant that derives sustenance from decaying wood or other plant parts, usually in association with a symbiotic fungus

scale A dry, flattened, papery body; sometimes also used as a term for a rudimentary leaf

scandent Climbing

scape The peduncle and rhachis of an inflorescence

seed A mature ovule containing an embryo and capable of germinating.

seed-coat The protective covering of a seed; also called testa

seedling A young plant raised from seed which has not yet flowered

sepal A segment of the calyx or outer whorl of the perianth

serrate With sharp, forward-pointing teeth

sessile Without a stalk, pedicel or petiole

sheath The base of a leaf or bract which embraces a bud or axis

shoot A term used by growers for a new growth

simple Undivided, unbranched; of one piece

spathulate Spatule-shaped or spoon-shaped

speciation The processes by which species evolve

species A taxonomic group of closely related plants all possessing a common set of characters which set them apart from another species.

spike A simple, unbranched inflorescence with sessile flowers

spur A slender, hollow projection from a floral segment (usually the labellum); spurs may be associated with nectaries

stamen The male part of a flower producing pollen, consisting of an anther and a filament

staminode A sterile stamen; often these are highly modified as in the column wings of *Thelymitra* and *Pterostylis*

stelidia A term used for the erect, horn-like or tooth-like column wings in *Bulbophyllum*

stellate Star-shaped or of star-like form

stem-clasping Enfolding a stem

sterile bract A bract which does not subtend a pedicel

stigma The enlarged, sticky area which terminates the pistil and is receptive to pollen and allows the pollen grains to germinate

stigmatic surface The sticky, receptive area of the stigma

style The slender part of the pistil which connects the stigma with the ovary; in orchids the style forms an indiscernible part of the column

subtend To support another structure or organ

subulate Awl-shaped; with a stiff point that tapers from base to apex

succulent Fleshy or juicy

sucker A shoot arising from the roots or the trunk below ground level

sulcate Furrowed

suture The markings, lines or ridges on an orchid capsule where it splits at maturity

sympodial A growth habit whereby each stem has limited growth and new shoots arise from the base of previous ones

sympatric Growing together

synonym An incorrect name which refers to a species

taxon A term used to describe any taxonomic group, for example, genus, species

taxonomy The classification of plants or animals

terete Round in cross-section and tapered

terminal The apex or end

terrestrial Growing in the ground

tetragonal Four-sided, as in the pseudobulbs of *Dendrobium tetragonum*

tomentose Densely covered with short, matted hairs

trilobed, trilobate With three lobes

truncate As if cut off square at the apex

tuber A thickened underground stem; not found in orchids (see tuberoid)

tuberoid A thickened, tuber-like root; these subterranean structures are found on many deciduous terrestrial orchids and are usually wrongly called tubers

umbel An inflorescence where the flowers radiate from a single point, as in *Bulbophyllum longiflorum*

undulate Wavy

unequal Of different sizes

unisexual Of one sex only; staminate (male) or pistillate (female)

variegated Where the basic colour of a leaf or petal is broken by areas of another colour, usually white, pale green or yellow

variety A taxonomic subgroup within a species used to differentiate variable populations

vegetation The whole plant communities of an area

vegetative Asexual development or propagation

vein The conducting tissue of leaves

velamen The layer(s) of thick, spongy cells which surrounds a root; common in epiphytic orchids

velutinous Covered with erect, stiff hairs

venation The pattern formed by veins

ventral On the lower side

viable Alive and able to germinate, as of seeds

villous Covered with long, soft hairs

viscid Very sticky or glutinous

viscidium A clearly defined sticky part of the rostellum which is removed together with the pollinia as a unit

viscous Very sticky

whorl Three or more segments (of leaves, flowers) in a circle at a node

wing A thin, membranous expansion of an organ; in orchid seeds it can refer to the dry, papery cells which surround the embryo; see also column wing

winged Having flat projections longitudinally along an axis

zygomorphic Assymetrical and irregular; a flower which cannot be divided equally in more than one plane

Bibliography

Ames, O. and Correll, D. S., *Orchids of Guatemala and Belize*, three parts bound as one, Dover Publications, Inc., New York, 1985.

Arditti, Joseph, ed., *Orchid Biology Reviews and Perspectives*, III, Cornell University Press, Ithaca, 1984.

Bailes, C., 'Pleione – A Neglected Genus', *American Orchid Society Bulletin*, 57 (1988): 493-9.

Barretto, G. D. and Young Saye, J. L., *Hong Kong Orchids*, Urban Council, Hong Kong, 1980.

Cribb, P., *Orchidaceae in Flora of Tropical East Africa*, A. A. Balkema, Rotterdam, 1984.

Correll, S. D., *Native Orchids of North America*, Chronica Botanica Coy, Massachussetts, 1950.

Curtis, Charles H., *Orchids, Their Description and Cultivation*, Putnam & Company, Ltd., London, 1950.

Darwin, C., *The Various Contrivances by which Orchids Are Fertilised By Insects*, second edition, revised, John Murray, London, 1882.

Dodson, C. H., '*Dressleria* and *Clowesia*: a New Genus and an Old One Revived in the Catasetinae (Orchidaceae)', *Selbyana*, 1 (1975): 130-7.

Dodson, C. H., 'Orchids of Ecuador: Stanhopea', *Selbyana*, 1 (1975): 114-29.

Dodson, C. H., 'Clarification of Some Nomenclature in the Genus *Stanhopea* (Orchidaceae)', *Selbyana*, 1 (1975): 46-55.

Dodson, C. H., 'The Catasetums (Orchidaceae) of Tapakuma, Guyana', *Selbyana*, 2 (1978): 159-68.

Dressler, R. L. & Pollard, G. E., *The Genus Encyclia in Mexico*, Associacion Mexicana de Orchideologia Mexico, 1976.

Dressler, R. L., *The Orchids, Natural History and Classification*, Harvard University Press, Cambridge, Massachussetts, 1981.

Dunsterville, G. C. K., *Introduction to the World of Orchids*, W. H. Allen, Switzerland, 1964.

Dunsterville, G. C. K., *Venezuelan Orchids Illustrated*, 1-6, 1959-1976.

Dunsterville, G. C. K., & Garay, L., *Orchids of Venezuela – an Illustrated Field Guide*, Botanical Museum, Harvard University, Massachussetts, 1979.

Du Puy, D. and Cribb, P., *The Genus Cymbidium*, Timber Press, USA, 1988.

Elliot W. Rodger and Jones, David L., *Encyclopaedia of Australian Plants & Suitable for Cultivation*, Vol I, Lothian Publishing Company Pty. Ltd. Melbourne, 1980.

Fowlie, J. A., *The Genus Lycaste*, Day Printing Corporation, California, 1970.

Fowlie, J. A., *The Genus Sophronitis*, Orchid Digest Reprint, (1972): 15-34.

Hawkes, A. D., *Encyclopaedia of Cultivated Orchids*, Faber and Faber Ltd., London, 1970.

Holttum, R. E., *A Revised Flora of Malaya*, Volume I, Orchids of Malaya, Government Printing Office, Singapore, 1964.

Jones, D. L., *Native Orchids Of Australia*, Reed Books Pty. Ltd. Sydney, 1988.

Lavarack, P. S. and Gray, B., *Tropical Orchids of Australia*, Thomas Nelson, Melbourne, 1985.

Luer, C. A., 'Icones Pleurothallidinarum', *Selbyana*, 3 (1976-77).

Millar, A, *Orchids of Papua New Guinea, an introduction*, Australian National University Press, Canberra, 1978.

Miranda, F. E., 'Catasetums in Brazil – Notes on Habitats and Culture', *American Orchid Society Bulletin*, 56 (1987): 473-82.

Marsh, R., 'Barkerias – the Shape of Things to Come', *American Orchid Society Bulletin*, 56 (1987): 581-7.

Northern, R. T., *Home Orchid Growing*, Third Edition, Van Nostrand Reinhold Company, New York, 1970.

Oplt, J., *Orchids*, Spring Books, The Hamlyn Publishing Group, London, 1970.

Pabst, G., *The Rupicolous Laelias*, Orchid Digest Reprint, (1984): 13-32.

Pessoa, C. & A., 'The Culture of the Brazilian *Sophronitis*', *American Orchid Society Bulletin*, 52 (1983): 1273-83.

Reinikka, M. A., *A History of the Orchid*, University of Miami Press, Florida, USA, 1972.

Schlechter, R., *The Orchids of German New Guinea*, translated by R. S. Rogers, H. J. Katz & J. T. Simmons, edited by D. F. Blaxell, Australian Orchid Foundation, Melbourne, 1982.

Schultes, R. E., *Native Orchids of Trinidad and Tobago*, Pergamon Press, New York, 1960.

Schultes, R. E. and Pease, A. S., *Generic Names Of Orchids, Their Origin and Meaning*, Academic Press, New York, 1963.

Schweinfurth, C., 'Orchids of Peru', *Fieldiana*: Botany 30(3), Chicago,1960.

Seidenfaden, G., 'Orchid Genera in Thailand', *Dansk Botanisk Arkiv* 1975 –.

Sheehan, T. and Sheehan, M., *Orchid Genera Illustrated*, Van Nostrand Reinhold Company, New York, 1979.

Soon, Dr. Teoh Eng, *Asian Orchids*, Times Books International, A. H. & A. W. Reed Pty. Ltd., Australia, 1980.

Sprunger, S., Ed., *Orchids from Curtis's Botanical Magazine*, Cambridge University Press, New York, 1986.

Stewart, J. and Campbell, Bob, *Orchids of Tropical Africa*, South Brunswick and New York, A. S. Barnes and Co. Inc., New Jersey, 1970.

Stewart, J., H. P. Linder, E. A. Schelpe and A. V. Hall, *Wild Orchids of Southern Africa*, Macmillan, Johannesburg, 1982.

Stewart, J., 'Portraits of Pretty Polystachyas', *American Orchid Society Bulletin*, 52 (1983): 1138-49.

Sweet, H., *The Genus Phalaenopsis*, Day Printing Corporation, California, 1980.

Van Der Pijl, L. and Dodson, C. H., *Orchid Flowers Their Pollination and Evolution*, University of Miami Press, USA, 1966.

Veitch, James and Sons, *A Manual of Orchidaceous Plants*, Volumes I & II, Reprint A. Asher & Co, Amsterdam, 1963.

Williams, L. O. & P. H. Allen, *Orchids of Panama*, Facsimile Reprint, Missouri Botanical Garden, 1980.

Index

D. S.C.t   11/86

Dictionnaire

des hérésies

**dans
l'Eglise
catholique**

Ouvrages du même auteur :

Dictionnaire initiatique, Belfond, Paris, 1970 (épuisé).
Dictionnaire (initiatique) des sciences occultes, de l'ésotérisme et des arts divinatoires, éd. rev. et augm. du *Dictionnaire initiatique,* Jean-Cyrille Godefroy-Sand, Paris, 1984.
Manual Diccionario de esoterismo, éd. Roca, Mexico, 1975.
Le Diable et la possession démoniaque, Belfond, Paris, 1975.
Le Diable et la possession démoniaque, Presses-Pocket, Paris, 1977.
La Gnose, une et multiple, éd. du Rocher, Paris, 1982.
Implosions, (Poèmes), éd. Caractères, Paris, 1980.
Les Prophéties de Paracelse, J.-C. Godefroy éd., Paris, 1982.
En collaboration avec Jean-Pierre Durand : *L'Ile Maurice aujourd'hui,* Jeune Afrique, Paris, 1983.

À paraître :
Les Heures bleues du Capricorne (Récits fantastiques et burlesques).
L'Homme-Équateur (Processus et métaphysique de la réalisation du Soi).